EXPLOITING REMOTELY SENSED IMAGERY

EXPLOITING REMOTELY SENSED IMAGERY

PROCEEDINGS OF
A ROYAL SOCIETY DISCUSSION MEETING
HELD ON 11 AND 12 MARCH 1987

ORGANIZED AND EDITED BY
K. A. BROWNING, F.R.S.,
B. J. CONWAY, J.-P. A. L. MULLER
AND D. J. STANLEY

LONDON
THE ROYAL SOCIETY
1988

Printed in Great Britain for the Royal Society
by the
University Press, Cambridge

ISBN 0 85403 349 1

First published in *Philosophical Transactions of the Royal Society of London*,
series A, volume 324 (no. 1579), pages 295–472

British Library Cataloguing in Publication Data

Exploiting remotely sensed imagery.
1. Remote sensing. Data processing
I. Browning, K. A. II. Royal Society
621.36′78

ISBN 0-85403-349-1

Published by the Royal Society
6 Carlton House Terrace, London SW1Y 5AG

PREFACE

Remotely sensed images from satellites and from airborne and ground-based sensors have become indispensable sources of data in environmental, commercial and military applications, yet our success in exploiting them is still patchy and limited. It is one thing to look at a satellite picture and gain a qualitative impression of agricultural patterns or weather systems; it is quite another to recognize and extract detailed quantitative information, reliably and in a form that allows it to be integrated and used with data from other sources. When such analyses have to be done under time constraints, or with large numbers of images, the difficulties are compounded and the need for fast, efficient and, if possible, automated methods becomes acute.

The importance and difficulty of exploiting remotely sensed imagery prompted us to organize a Discussion Meeting which was held at the Royal Society on 11 and 12 March 1987. The first half of the meeting concentrated on how images are already being put to work: the use of interactive computer-based display systems in fields such as oceanography and weather forecasting, the tasks that become practicable with very powerful computers, and the problems of incorporating image-derived data into comprehensive geographic information systems. On the second day attention was transferred to progress on new techniques, notably the development of intelligent knowledge based systems capable of some of the judgemental functions currently performed by human beings. Throughout the meeting the emphasis was on the use of remotely sensed imagery in real applications of practical importance, rather than on image-processing as an end in itself.

The development and deployment of new instruments continues, offering an ever richer choice of image data but presenting even greater challenges to users. The papers in this volume show how remotely sensed imagery is already proving invaluable in demanding practical tasks, but point to the new work which must bear fruit before we can fully exploit the wealth of data now becoming available.

We thank the staff of the Royal Society for their assistance, particularly Miss Christine Johnson for her help in arranging the meeting, and Ms Mary Healy for preparing this volume for publication.

January 1988

K. A. Browning
B. J. Conway
J.-P. A. L. Muller
D. J. Stanley

CONTENTS

[19 plates; 1 pullout]

Phil. Trans. R. Soc. Lond. A **324**, 297–298 (1988)
Printed in Great Britain

Chairman's introduction

By K. A. Browning, F.R.S.

Meteorological Office, London Road, Bracknell, Berkshire, RG12 2SZ, U.K.

This meeting is about ways in which to use imagery, mainly but not entirely satellite imagery. What spurred Dr B. J. Conway, Dr J.-P. Muller, Dr D. Stanley and me into organizing the meeting is a concern about the lag between initial availability of satellite data and the eventual full exploitation of the data. As a meteorologist I am rather proud of the way in which we are reaping benefits from space techniques. Yet I am acutely aware that the first cloud pictures from a weather satellite were available more than a quarter of a century ago, whereas it is only quite recently that we have begun to develop procedures for using the imagery quantitatively and to adjust our operational practices to capitalize on it. This kind of delay is not unique to meteorology. The question is, why in general there should be such a delay?

The first reason for the delay in exploiting imagery is the disparity between the sums of money spent on launching satellites and building the instruments and the comparatively small sums of money available for ground facilities and the development of ways in which to use the data.

The second reason for the delay in exploitation is that digital imagery is difficult to handle. For a start, the amounts of data are prodigious. As that pioneer of satellite techniques, Verner Suomi, has so vividly described it, handling the flow of data associated with satellite imagery is like trying to drink from a fire hydrant. A necessary development has of course been the advances in computer hardware but that is not the whole story; the problem is that there is often a complex relation between the wanted information and what the satellite actually senses or resolves. There is a need to interpret the imagery in the light of a number of different strands of evidence. This evidence is in the form of *in situ* measurements, so-called 'ground truth' (but not always very true or representative), and also prior knowledge or understanding of aspects of the scene. These strands of evidence tend to be fragmentary and equivocal. Human judgement is needed to weigh the evidence.

It is obviously important to have a good man–machine interface if the human operator is to exercise his judgement effectively within the context of an otherwise highly automated image-handling system. In this meeting you will hear about the development of interactive computer display systems which act as fast, willing slaves, freeing their human masters for higher-level interpretive and decision-making tasks. You will also be hearing about attempts to learn how the human experts arrive at their judgements, and to find out what scope there is for developing systems with the intelligence to perform autonomously some of the higher functions that the human operator presently performs.

The third reason for the delay in exploiting imagery is the communications gulf between the image-processing specialists on the one hand and the potential users on the other. This meeting is an attempt to bridge that gulf. We have brought together people with hands-on experience in the use of imagery in various fields: in meteorology, in oceanography, and in the measurement of land use and man-made structures. My hope is that by sharing experiences in the use of imagery in different fields we might see how techniques developed for one application may be used in another. As our last speaker will demonstrate, we do not have to depend solely on experience in remote sensing to gain insight into ways of interpreting remotely sensed imagery.

The way this discussion meeting has been structured is that the first day's papers deal mainly with the present situation. We shall hear about some of the currently available imagery, with concrete examples of how remotely sensed images are already being exploited. Then on the second day we shall hear about some of the image processing research that holds promise for the future; some first steps upon a long road.

Phil. Trans. R. Soc. Lond. A **324**, 299–315 (1988)
Printed in Great Britain

Weather forecasting by interactive analysis of radar and satellite imagery

BY B. J. CONWAY AND K. A. BROWNING, F.R.S.
Meteorological Office, London Road, Bracknell, Berkshire RG12 2SZ, U.K.

[Plates 1 and 2]

At the Meteorological Office, Bracknell, quantitative rainfall maps from a network of ground-based radars, augmented by cloud images from *Meteosat*, are used to produce analyses and very-short-period forecasts of precipitation. These remotely sensed images provide the only way of presenting the current weather situation quickly enough and with the required spatial resolution and areal coverage. The processing of the radar and satellite data is highly automated, but there are some tasks that require judgements based upon many strands of information and an understanding of meteorological processes. To this end, forecasters use a specially developed display system to interact with the imagery. The facilities for interacting with the pictures have been optimized so that the forecaster, who is kept very busy in active weather situations, can keep pace with the flow of real-time data. Even so, as more radars are added to the network, ways must be found of reducing the burden of the forecaster's interpretive and judgemental functions by automating some of them and making others easier to perform.

1. INTRODUCTION

Remotely sensed images from a ground-based radar network and from the European geostationary satellite, *Meteosat*, are being used in combination as a routine part of the U.K. Meteorological Office's precipitation forecasting operation (Browning & Collier 1982). Every half hour, quality-controlled current rainfall maps and a series of extrapolation forecasts for a few hours ahead are generated for a limited area embracing England and Wales. Such information is required for flood prediction and as an input to detailed forecasts that are of value to a variety of users including farmers, the construction and leisure industries, and aviation. Some of these products are already being disseminated; others are still under development.

No other type of observing system can provide data with the spatial and temporal resolution required, but the image data, although our primary source, are not sufficient by themselves. Pure image-processing, however sophisticated, cannot compensate for all the ambiguities and errors that occur in these data, which must instead be interpreted in the light of knowledge from a variety of sources, including conventional observations, numerical model predictions and conceptual models of mesoscale atmospheric processes (Browning 1986). Weather forecasters are used to combining different data sources to construct forecasts, but they need new methods if they are to be able to work with images as easily as they have traditionally worked with isolated observations and line-drawn charts.

Appropriate presentation of the images is essential if the human operator is to stand any chance of recognizing the important relations contained within the pictures. It is particularly important to be able to view images in relation to others and as members of sequences; organization concealed in a single picture becomes instantly apparent when images are viewed as a rapid temporal sequence. All the images, both radar and satellite, must be presented on a common projection, and means of alternating between images and of combining one picture with another must be provided. Operations on single images, to delete, add or highlight selected features, must be available so that the forecaster can not only view the images but modify them. It must be remembered that images are not just input data used by the forecaster; they are also the natural means of presenting the final product to the customer.

The value added to the raw-image data by careful interpretation would be quickly lost if this resulted in a large delay in distributing the products to users; the total forecast period is only a few hours and the value of current rainfall maps decays quickly. None of the tools in a traditional forecast office equips the forecaster to handle images in the ways required, much less to do it quickly enough to be of any practical use.

The Meteorological Office's answer to these demanding image-handling requirements has been the development of a special-purpose interactive display system called FRONTIERS, which allows an experienced forecaster to examine, modify and combine the images, finally transmitting them to users throughout the country. FRONTIERS provides a range of facilities to help the forecaster and relieve him of routine, low-level tasks, while still leaving him in control, with the flexibility to modify the images as he wishes and to decide which functions to perform. With such a system it is possible to analyse and correct the radar rainfall images within minutes of receiving them and very shortly afterwards to have generated and distributed a sequence of extrapolation forecasts. The immediacy of this process and the dominance of the current situation in these very short term forecasts are encapsulated in the term 'nowcasting' which has been coined to describe operations such as these.

2. The FRONTIERS system

FRONTIERS was designed and built by Logica to a Meteorological Office functional specification and delivered in initial form in 1983. Experiments at the Meteorological Office's Radar Research Laboratory had already shown the kinds of task that the forecaster would need to perform when analysing the images and constructing the forecast, and work at other laboratories (Suomi *et al.* 1983) had demonstrated the use of interactive display systems for meteorological analysis, so it was possible to specify with confidence the facilities needed (Browning 1979). FRONTIERS is truly interactive; the forecaster uses it to modify images as well as to view them, and it is designed to put him in direct contact with the images without being conscious of the computer as an intermediary.

FRONTIERS comprises an interactive work station (figure 1) supported by a DEC VAX 11/750 computer. Within the work station the images are presented on high-performance colour displays, one of which is fitted with a transparent touch-sensitive panel so that the forecaster can operate on the images simply by pointing to features of interest with his finger. This colour monitor is flanked by menu screens on which are presented the options available to the forecaster at that point. Again, these are fitted with touch-screens so that selections are made by pointing to the desired options; there is never any need to remember file-names or computer commands or to type these at a keyboard. Other input devices, a joystick and a

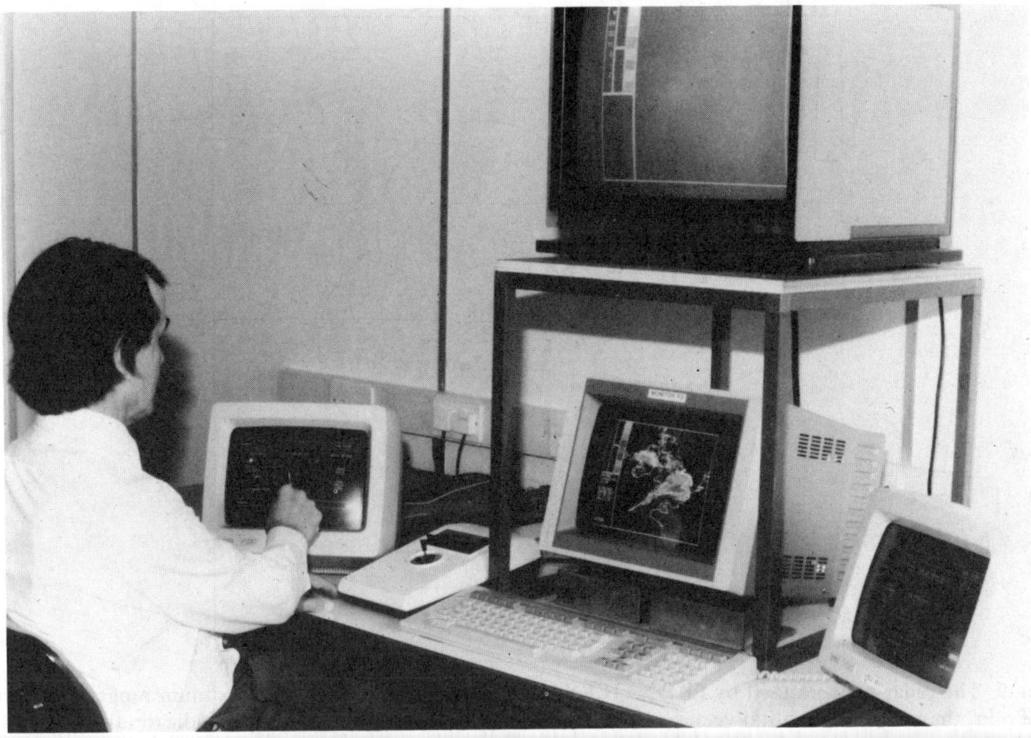

FIGURE 1. The FRONTIERS work station.

graphics tablet, are available for those functions to which they are well suited. Great attention is paid to making the forecaster at ease with the equipment: the system remembers which forecasters are left handed and swaps the menu panels, and every forecaster has his own preferred colour table in which the images are presented. The central idea is that the forecaster can give his entire attention to the job in hand, rather than to operating the equipment.

The FRONTIERS input data consist of quantitative rainfall maps from a network of ground-based radars and images from the European geostationary satellite *Meteosat*. Rainfall measurements in the U.K. network (figure 2) are made by the 'radar reflectivity' method; the rainfall rate is deduced from the strength of the return signal. Each radar site transmits its measurements as an array of 84 pixels × 84 pixels, each pixel showing the instantaneous rainfall rate averaged over a 5 km square. Readings from four or five telemetered rain gauges are sent with each radar image for calibration and as an independent check on the radar data. Measurements from all the sites are received every 15 min, but only the half-hourly pictures are used directly in FRONTIERS products; the intermediate ones are available to various users, but are used within FRONTIERS itself just to improve the continuity of replay sequences.

Digital images from *Meteosat* are received locally at half-hourly intervals. *Meteosat* scans the earth in three spectral channels: visible (0.4–1.1 μm), water vapour (5.7–7.1 μm) and infrared (10.5–12.5 μm). The horizontal resolution in the region of the southern U.K. is about 6 km × 3 km in the visible channel and 12 km × 6 km in the other two, but the images are sampled onto the same 5 km × 5 km resolution National Grid projection used for the radar images so that all the image products are readily compared. Although *Meteosat* images have poorer resolution

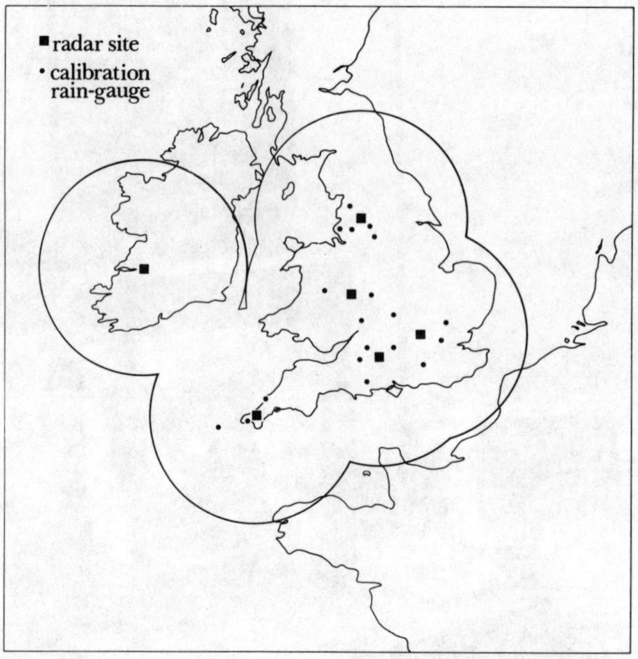

FIGURE 2. The radar network used by FRONTIERS in 1987. The circles show the maximum range for detection of rain; the range for quantitative measurements is smaller. The frame of the picture indicates the extent of the *Meteosat* image area used by FRONTIERS.

and are available in fewer spectral channels than the images from the NOAA polar-orbiting meteorological satellites, they have the overriding advantage for nowcasting of being repeated every 30 min rather than at intervals of several hours.

3. The FRONTIERS operational cycle

FRONTIERS operates on a half-hourly cycle synchronized with the transmissions from *Meteosat*. In each cycle three major tasks are performed: a quality-controlled radar composite is produced, *Meteosat* images are used to extend the area of coverage and very-short-period rainfall forecasts are generated by extrapolation. Each of these tasks is divided into a sequence of smaller steps. Here we concentrate on a few selected operations which show the essential features of FRONTIERS and how it allows a forecaster to interpret and modify the image data in the light of his understanding of the weather situation, using a mixture of objective and subjective processes.

(*a*) *Radar analysis stage*

In this first stage the forecaster attempts to make the best possible interpretation of the latest radar network rainfall image, using the various data sources at his disposal to build a picture of the weather situation, of which the precipitation is one facet. Many of the errors in the radar images are not limitations intrinsic to radar, but are meteorological in origin, so that attempts to correct them are inseparable from an understanding of the weather systems being observed. Figure 3 shows some of the more important sources of error.

The first operation in the FRONTIERS cycle is to assemble a composite picture from the individual radar site data. Some corrections are applied automatically at the radar sites but

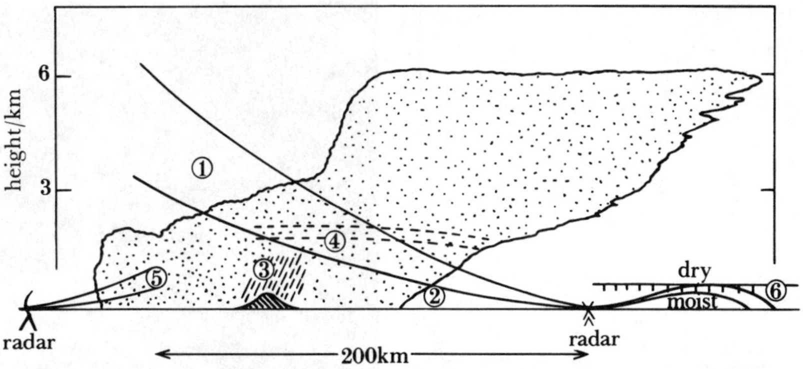

FIGURE 3. Cross section through an area of frontal precipitation showing six sources of error in the radar measurement of surface rainfall intensity: (1) radar beam overshooting shallow precipitation at long range, (2) low-level evaporation beneath the radar beam, (3) orographic enhancement above hills which goes undetected beneath the radar beam, (4) anomalously high radar signal from melting snow (the 'bright band'), (5) underestimation of the intensity of drizzle because of the absence of large droplets and (6) radar beam refracted by a strong vertical humidity gradient causing it to intercept land or sea (anomalous propagation).

most of these are removed to provide a fresh start for the more comprehensive FRONTIERS correction procedure. The operator's main task in overseeing the construction of the composite is to recognize gross errors caused by radar malfunctions and, if necessary, to remove completely the data from the offending radar. Such faults are usually instantly recognized by the operator when displayed as a rainfall map, though they are not easily identified by the radar site computer.

The next task is to try to identify and remove spurious radar echoes, that is any echo which does not correspond to rain at the surface. These include echoes produced by backscatter from the surface (ground and sea clutter) in conditions of anomalous beam propagation, and those resulting from rain which evaporates before reaching the ground. To do this job properly the operator must take into account information from a variety of sources, in addition to that in the radar images themselves.

It is difficult to recognize ground and sea clutter echoes in a single image, particularly if they are not at habitual clutter locations. Their appearance is often indistinguishable from that of genuine rainfall. The operator therefore uses FRONTIERS to replay repeatedly a series of images as a 'movie loop'. Most spurious echoes caused by ground returns then show up clearly because their stillness or erratic movement contrasts strongly with the organized movement of real rain systems. *Meteosat* cloud imagery is sometimes useful in the identification of spurious echoes. If no cloud can be detected in the region of the echo then it is unlikely to be rain. Spurious echoes that happen to occur in an overlap region between two radars may also be revealed if the image from the appropriate radar is alternated rapidly with the composite image with that radar removed; a particular ground clutter echo is usually seen by only one radar so it will flash on and off as the picture alternates. There is some prospect that spurious echoes from ground returns can be trapped by specialized hardware at the radar receiver, by using the temporal characteristics of the signals, but sea clutter is less easy to distinguish by such means and is likely to remain a task for FRONTIERS.

For most practical purposes we are interested only in rain that reaches the ground, so it is important to try to detect instances where rain detected aloft is evaporating in a low-level dry layer (figure 4). In doing this the operator will use many strands of evidence from different

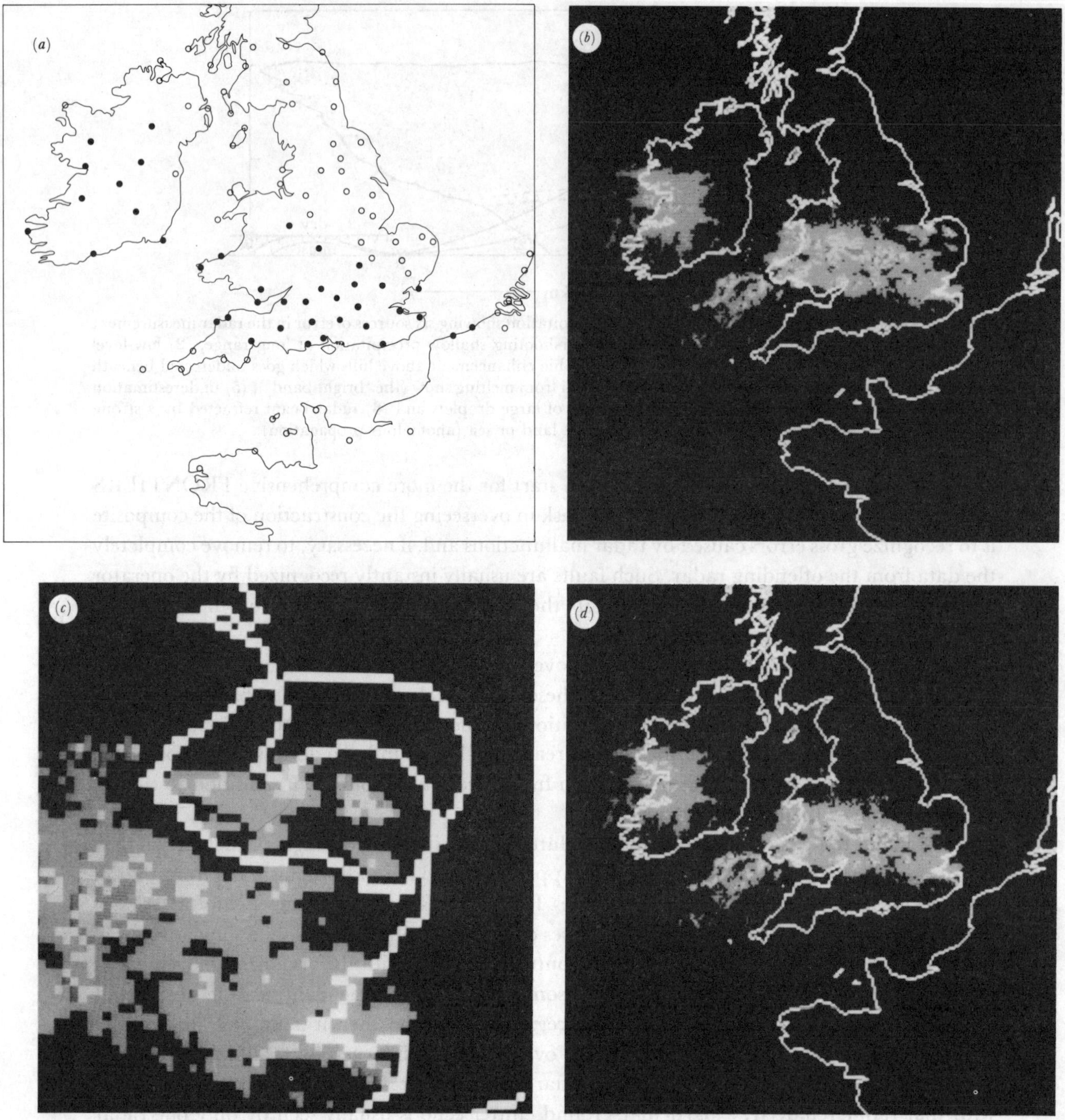

FIGURE 4. Correcting for low-level evaporation of rain in the FRONTIERS radar analysis stage. (*a*) Surface observations (from conventional meteorological stations) of rain (●) and no rain (○) at 03h00 GMT on 9 February 1987. (*b*) Radar composite for the same time as in (*a*). Rain is detected aloft, south of the Wash, by the London radar but is evaporating before reaching the ground. (*c*) The FRONTIERS operator zooms in on the evaporating rain and marks it for deletion by drawing a boundary round it on the screen. (*d*) The radar composite of (*b*) with the evaporating rain removed. The radar is missing some light rain and drizzle near the south coast because of the small drop sizes. The operator gets a chance later in the FRONTIERS cycle to insert rainfall here.

sources. He will be helped in this by the fact that FRONTIERS is located in the Central Forecasting Office at Bracknell and is an integral part of a much larger forecasting operation. The operator will be aware of the synoptic situation (the large-scale pattern of weather) and will relate that to his knowledge of where the conditions for low-level evaporation are often to be found, for example just ahead of a warm front. Having identified the regions in which the effect is likely he will first see how far they are from the radar sites and, from this, whether the atmosphere is being sampled a long way from the ground. Unless he is convinced that rain is being measured low enough to be sure of reaching the surface he will go on to seek more specific evidence. Radiosonde ascents, which show the vertical temperature and humidity profiles of the atmosphere, may provide direct confirmation of the presence of a dry layer but will not normally allow precise definition of the area affected. Numerical forecast models will often provide a clearer indication of the pattern of humidity. The most useful information, however, is that from conventional surface observations and from telemetered raingauges. The relative importance of these different sources is not fixed, but depends on how they relate in space and time to the feature being considered; it is up to the operator to weigh them accordingly.

The time arrives when the operator must decide which pixels must be deleted as not representing surface rainfall. It is unlikely that the observational evidence alone will have been sufficient to permit an unequivocal delineation of the region of low-level evaporation and so the operator must use his judgement. By now he will have built up a three-dimensional picture of the atmosphere in the suspect region based on a mental model of the synoptic feature, fitted to the available observations in this specific case. He now uses the FRONTIERS facilities to draw boundaries on the screen within which echoes should be deleted, or alternatively to zoom in and delete individual pixels. The machine places him in direct contact with the images and makes it easy to modify them and to change his mind as often as time permits. Thus his full attention can be given to the changes he wishes to make rather than to the mechanics of the operation. Only when he is satisfied does he instruct the machine to implement the changes permanently and move on to the next step.

Four further steps remain before the FRONTIERS radar analysis stage is complete. If the presence of enhanced echo from melting snow (the so-called bright band) has been detected at any radar site the operator must examine corrections computed by FRONTIERS and decide whether to apply them. He must define areas of different types of rainfall, so that appropriate calibration factors can be applied and so that the range at which the beam will overshoot the rain and be of limited further use can be calculated. He must then look for evidence of individual radar calibration failures, where the sensitivity of a particular radar has changed significantly, and try to correct this. Lastly, he must decide whether intensification of rain over hills (so-called orographic enhancement) is likely to be important, and if so where. Orographic enhancement is a low-level phenomenon and often occurs beneath the radar beam. Thus it is necessary for the operator to instruct FRONTIERS to calculate and apply the appropriate correction factors.

In all of this the operator will be examining the radar measurements critically in the light of other information, just as he did for deletion of spurious echoes, and using FRONTIERS to implement modifications, calculate corrections and show quickly the results of his decisions. The operator has, at most, 17 min to complete all the steps listed above; if he has not finished by then FRONTIERS will take control and finish the radar analysis automatically, applying the corrections used in the previous cycle. If the operator can finish sooner he is rewarded by getting extra time for the satellite analysis step which follows, but if he takes less than 12 min he has to wait for the satellite images to arrive. This shows the extremely tight scheduling of the FRONTIERS cycle and the paramount importance in the design of the system of an efficient man–machine interface.

(b) *Satellite analysis stage*

In the second major part of its 30 min cycle FRONTIERS is used to estimate, from the satellite cloud imagery, where rain is likely to be falling outside the radar area, so that a less fragmentary glimpse of the overall pattern can be obtained and so that some warning can be given of approaching rain systems.

The *Meteosat* images do not show rain directly, so its presence must be inferred from cloud images (figure 5, plate 1). FRONTIERS allows this to be done by a mixture of objective and subjective techniques. Different information is available from the different spectral channels: the infrared image shows the temperature, and therefore the height, of the cloud tops, whereas the brightness of the visible image indicates the thickness of the cloud. Both are important in assessing the likelihood of rain falling from the cloud. At night, however, the infrared image must be used alone.

The preferred method of relating the cloud information to the rainfall pattern is to correlate both the visible and infrared images with the rainfall measurements within the radar area, and then to assume that this correspondence holds in other areas as well (Lovejoy & Austin 1979). The range of possible pixel values in each channel is divided into 16 equally sized classes so that the cloud at a given location will correspond to one of $16 \times 16 = 256$ possible visible–infrared classes. Each cell within the radar area is examined in turn and 'rain' or 'no rain' is entered against the cloud class at that location. At the end of this process we obtain, for each cloud class present in the data, the fraction of occurrences corresponding to rain, which can be taken as a measure of the probability of rain falling from that class of cloud.

The relation thus obtained is then applied, in reverse, to the whole area. All pixels in the cloud category with the highest rain probability are assigned as rain, then all those with the next highest probability, and so on, until the total number of pixels of inferred rain within the radar area is as close as possible to the number of radar rain pixels. This method prevents grossly inaccurate estimates of the amount of rain but, because the relation is imperfect, the pattern of the inferred rainfall will not exactly match that of the radar rain, and the degree of mismatch, both within the radar area and at its boundary, is used by the operator to gain a subjective estimate of the success of the operation and the acceptability of the correlation product.

FRONTIERS gives the operator the option of correlating rainfall against either the visible

---

DESCRIPTION OF PLATE 1

FIGURE 5. Derivation of an extended area rainfall field from *Meteosat* cloud images. (*a*) Radar composite for 13h00 GMT on 4 August 1985, showing observed rainfall rates after correction for errors in the FRONTIERS radar analysis stage. Green shows lightest rain (less than 1 mm h^{-1}), blue shows heaviest (8–16 mm h^{-1}). (*b*) *Meteosat* visible (0.4–1.1 μm) image for the same time as in (*a*). The whitest areas correspond to the brightest and therefore the thickest cloud. (*c*) *Meteosat* infrared (10.5–12.5 μm) image for the same time as in (*a*). Colours show the temperatures and therefore the heights of the cloud tops. The coldest cloud tops in the picture are indicated by the orange patches over eastern England and Northern Ireland. Very dark areas are warm cloud or surface. (*d*) Field of rainfall at 13h00 on 4 August 1985, inferred from the *Meteosat* visible and infrared images by correlating them with the radar rainfall pattern within the radar area. This field shows only the extent, not the intensity, of the rain.

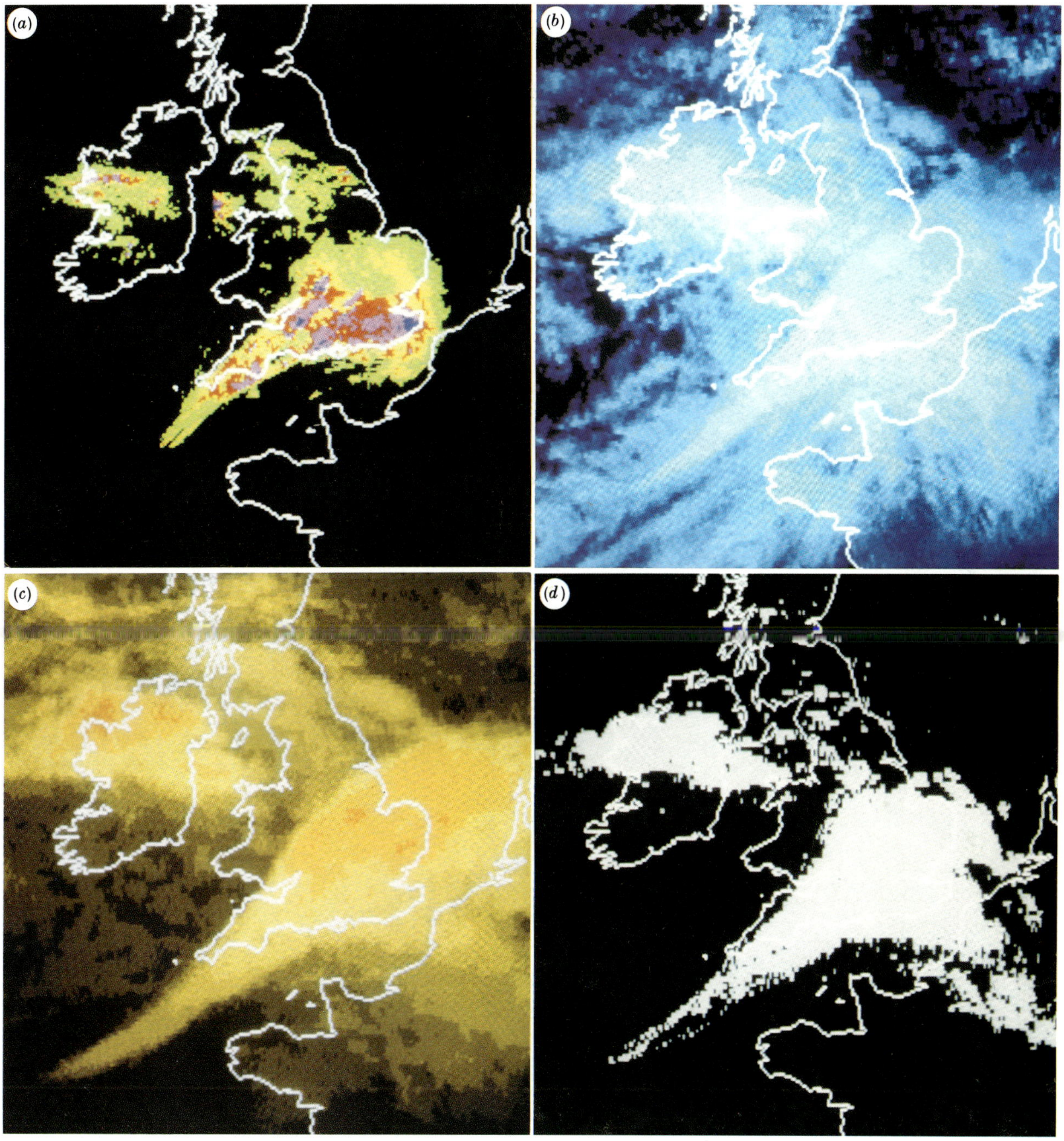

Figure 5. For description see opposite.

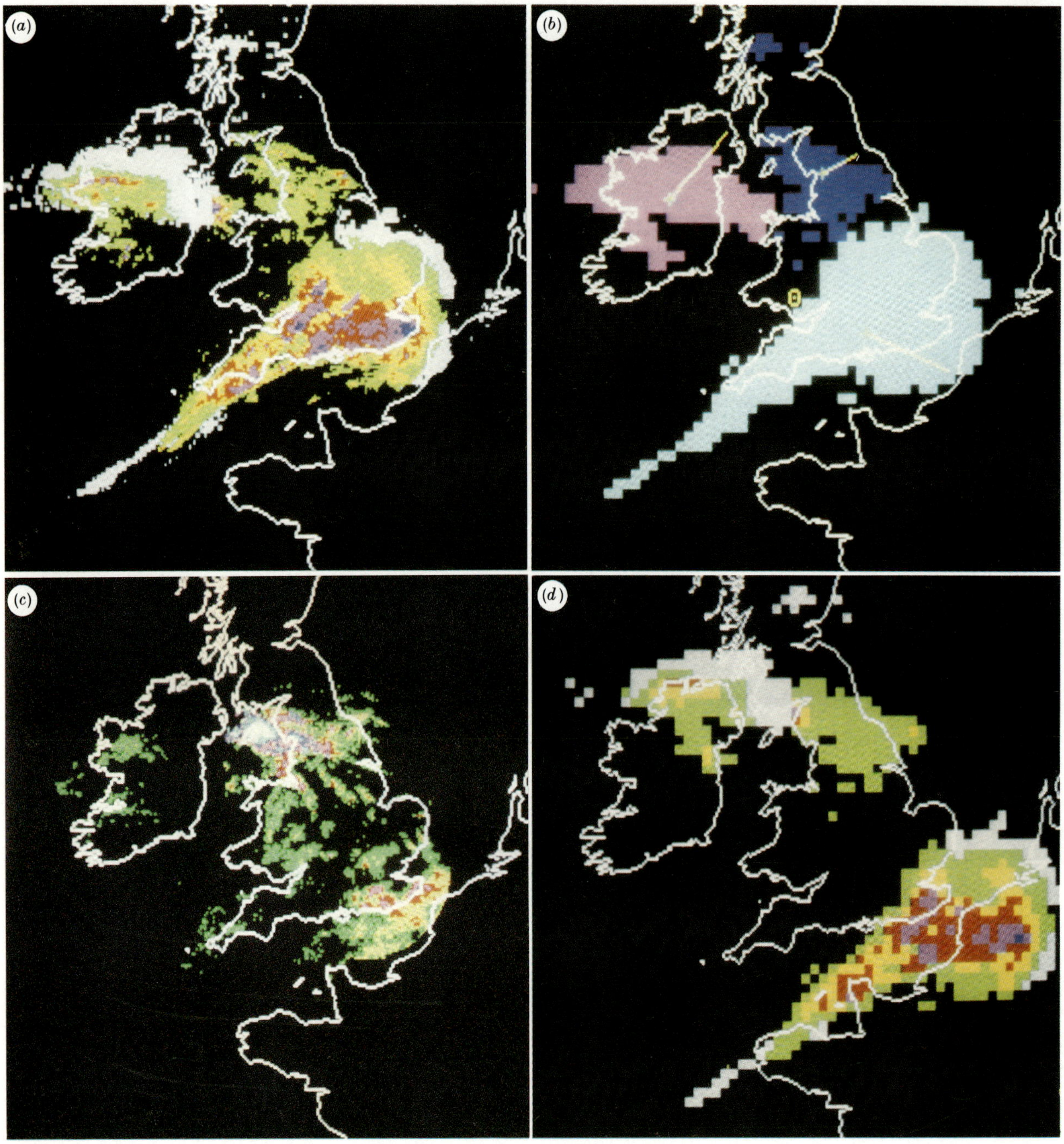

FIGURE 6. For description see opposite.

image, the infrared or, as described above, the two combined. Normally he would try the visible–infrared combination first, provided that both images were available, but if dissatisfied with the result he might try the other options. The choices available are presented on a touch-sensitive menu screen, so that exploring the different possibilities is quick and easy.

In most cases one or other of the correlation products will provide a reasonable approximation to the desired rainfall field, but the method is subject to a number of problems.

(i) If there is very little rain in the radar area, the correlation will be unreliable. This is countered by using a continuously evolving correlation table which consists partly of the present correlation and partly of the table used last time. If this fails, a fixed 'universal' table is available.

(ii) If the predominant types of rainfall inside and outside the radar area are different then the assumption that the same relation between cloud class and rain holds in both regions may be invalid.

(iii) The method is seriously weakened at night when only the infrared image is available, because the cloud-top temperature does not on its own bear a close or reliable relation to the presence or absence of rain, except for certain types of rain (convective storms).

(iv) A small misregistration of the satellite image with respect to the radar rainfall map can have a marked effect on the correlation result if the cloud and rainfall patterns contain much fine structure.

The operator must decide whether to accept one or other of the correlation products. He usually does this by comparing the satellite derived field with the radar analysis product, using a FRONTIERS facility which allows the correlation rainfall field to be displayed with the radar field embedded. By using FRONTIERS to alternate rapidly between this and the complete correlation field, the operator can obtain a clear impression of how well the correlation product agrees with the radar measurements and whether the pattern outside the radar area is a realistic extension of what he sees inside it. In this he will also be guided by his knowledge of the synoptic situation and his expectation of the organization, within the large-scale pattern, of mesoscale features such as fronts and rainbands.

The operator may occasionally accept the correlation product unaltered, but more often he will decide to modify it either by deleting features he regards as spurious or by adding areas of rain where he believes that the correlation field contains unrealistic gaps or discontinuities. FRONTIERS provides the means to do this by direct interaction with the image, with pan and zoom facilities permitting accurate work on magnified areas. As usual with FRONTIERS, mistakes are easily cancelled *en route* to constructing a final product.

Description of plate 2

Figure 6. Generation of a rainfall forecast. (*a*) The radar–satellite combination rainfall field formed by adding the satellite-derived rainfall field of figure 5*d* to the quantitative radar rainfall field of figure 5*a*. The satellite-derived rainfall (shown in white) has been selectively trimmed by the FRONTIERS operator. (*b*) The rainfall pattern of figure 6*a* divided into clusters that will move independently in the forecast. Their velocities are indicated by the white and yellow vectors at their centroids. (*c*) The rainfall pattern 3 h later, at 16h00 as observed within the area of coverage of the radar network. (*d*) The 3 h extrapolation forecast for 16h00. The heavy rainfall in SE England is well forecast, as is the extent of the rain in northern England and the Scottish border region. The very heavy rain over the Irish Sea would have been missed almost entirely if the satellite data had not been available to bridge a gap in the radar coverage (figures 5*a* and 6*a*). The light rain over the Midlands and Wales developed *in situ* during the forecast period and so could not be forecast by extrapolation techniques. On this occasion the forecast actually gives better agreement with the surface observations in Northern Ireland than does figure 6*c* because the radar at Shannon (SW Ireland) is missing some rain in that area at 16h00 GMT.

Sometimes the operator will decide that none of the available correlation fields has represented the rainfall field well enough to be of any use, even as a starting point. Another option available to him is to choose a satellite image and to 'slice' it so that only a chosen range of pixel values is made visible on the screen, pixels of any other value being suppressed. By varying the size and position of this chosen range, the operator can selectively display features within the satellite picture, picking out clouds of a particular range of brightness or temperature. The operator controls the slicing by means of a joystick while watching the image evolve on the screen, so that from a very large number of initial possibilities he can quickly converge on the most realistic product. Experience has shown that a satisfactory product is more likely to be obtained by slicing the visible image rather than the infrared, so it is less easy to obtain good results by this method at night. As with the correlation field, the operator will often choose to make alterations to the sliced image, adding or deleting rain, until he is happy with the product.

As a last resort, if no satisfactory pattern can be obtained either from the correlation process or from slicing an image, the operator has the means to draw in the rainfall pattern from scratch, freehand, to produce what he considers to be a realistic pattern, agreeing with the radar data and with the current weather situation as far as he is able to determine it from the other sources at his disposal. This method produces a coarser rainfall field than the other techniques and is used on its own only when they fail. Nevertheless, it does provide a way of giving users an impression of the likely overall pattern of precipitation and a means of avoiding unnatural discontinuities associated with data gaps being advected into the radar area.

This completes the second stage of the FRONTIERS cycle, the extension of the area of coverage. The system provides the means to process this stage by a range of techniques, from the purely objective to the purely subjective, depending on the details of the particular case. The product forms the starting point for the next stage.

(*c*) *Forecast stage*

The last stage in the FRONTIERS cycle is the generation, by extrapolation of recent movement, of a forecast sequence showing the expected distribution of rainfall for a few hours ahead. The period of validity of such forecasts varies with the type of weather situation, but for the sake of delivering a uniform product, likely in most cases to show some skill throughout the forecast period, it has been decided for the present to generate a series of half-hourly forecasts stretching a fixed 3 h period ahead.

The extended area composite from the satellite analysis stage (figure 6*a*, plate 2) is first reduced to 20 km horizontal resolution by averaging neighbouring pixels. This was originally done to save computing time later in the forecast, but it has undesirable effects such as causing jerky motion of the rainfall patterns in the forecast sequence and reducing the contrast in the rainfall intensity field. We therefore plan to revert to using 5 km resolution throughout, and solve the problem of computing power in other ways (see §4).

Rarely will a single velocity vector adequately describe the motion of the whole rainfall field. The way this is taken into account in FRONTIERS is for the operator to divide the rainfall pattern into a number of clusters which he believes will maintain their identity during the forecast period (figure 6*b*, plate 2). He does this by simply drawing the boundaries between clusters on the screen using his finger. In choosing how to dissect the image the operator uses not only his knowledge of the pattern of weather producing the precipitation, but also the evolution of the pattern in the recent past, as revealed by replay sequences of the radar and satellite images.

Having selected the rainfall clusters that will populate his forecast, the operator must assign them appropriate velocities. Although a general impression of movement within an evolving weather system is easily obtained, measurement of the precise velocities of chosen features proves to be much more difficult. So far, our attempts to develop totally objective procedures have been less successful than our efforts to optimize the implementation of subjective methods.

One subjective technique is to calculate the velocities from the difference in position of the features at two known instants. Problems arise because if a short time interval is chosen, measurement of the small movement that has occurred is inaccurate, whereas if a longer interval is used to give a larger movement, changes in the shapes of the features will often make identification of corresponding points uncertain. Although FRONTIERS allows the operator to define velocities in the above manner, a much preferred technique which has been specially implemented in FRONTIERS for this purpose is the 'lagrangian replay' facility. This exploits the human operator's ability to separate a chosen feature from a confusing background and then follow it as both it and the background move and distort. This is possible provided he can see the sequence as a continuously moving picture, not as a series of disconnected 'stills'.

The operator first selects the type of image to use (radar-rainfall or satellite) and within it a feature whose velocity he wishes to measure. He chooses a short replay sequence, terminating on the latest image, throughout which the feature can be identified. He zooms in, centring on the feature, and touches corresponding points on the first and last frames so that the system can compute a 'first-guess' velocity. The sequence of images is then repeatedly replayed, with the system subtracting the first-guess velocity from the entire field so that the chosen feature appears roughly stationary. The operator then adjusts the compensating velocity by means of the joystick, while the replay sequence is running, until the feature's residual velocity is minimized, and in this way a reliable estimate of the required velocity is obtained. It is found that the lagrangian replay is the method that gives the most consistent results when the same case is processed by several different operators. Even when a feature significantly changes during a replay sequence, thereby rendering automatic procedures unreliable, there tends to be good agreement among different observers about when it appears stationary. The lagrangian replay thus proves to be a powerful subjective technique for measuring velocities in complex evolving systems.

When a number of velocities, characteristic of the identified separate components of the rainfall system, have been measured, they are assigned to the appropriate clusters simply by the operator touching the appropriate velocity on a displayed list and then touching the corresponding cluster. A particular velocity may be assigned to more than one cluster. After that the generation of the forecast sequence is largely automatic, although the operator has some choice about how orographic corrections should be applied in the forecast fields and whether he wishes to define multiplicative factors (in the light of other evidence) by means of which rainfall in the various clusters can be made to increase or decay with time.

Many aspects of the extrapolation forecast are undeniably crude. Clusters move in straight lines at constant speed without changing shape, and only the simplest representation of development and decay is included. Because the technique only depicts a continuation of observed recent trends, the development of completely new rainfall areas can never appear in the forecast. Nevertheless, in many situations, particularly in the frontal patterns which dominate rainfall events in the U.K., the method gives a good indication of the immediate movement of the rainfall pattern and constitutes useful precipitation forecast guidance in the very short term (figure 6*c*, *d*, plate 2).

4. Future developments

FRONTIERS is now undergoing operational trials in the Central Forecasting Office at Bracknell. The quality-controlled radar composite map is being routinely disseminated to users around the country but other FRONTIERS products await the availability of improved display and transmission facilities which the Meteorological Office plans to install over the next few years (Cluley & Hills 1988). In the meantime, FRONTIERS extended rainfall maps and forecasts are being used for guidance centrally.

Development of FRONTIERS continues. Even if the system were perfect in itself, the fact that it is part of a changing forecasting environment would mean that it must evolve to take advantage of new sources of data and to deliver products meeting practical needs. Another spur to development is the need to increase the throughput of the system. The operator is already kept very busy in active weather situations. With the imminent expansion of the U.K. radar network and the possibility of the use of data from continental radars, he will be extremely hard pressed to keep up with the flow of real-time data.

The further development of FRONTIERS will be approached in three principal ways: (*a*) through use of improved but standard computing techniques, (*b*) through interaction with a dynamical weather-prediction model and (*c*) through the application of expert-system techniques.

(*a*) *Use of improved but standard computing techniques*

Improvements will be made to the speed of the FRONTIERS computer system and to the ergonomics of some of the facilities. The machine is annoyingly slow when responding to some commands and, although it may be doing a lot of work at the time, this is not usually apparent to the operator, who is conscious only of having to wait before proceeding. These difficulties can be largely overcome by straightforward computing techniques (improved software efficiency, more powerful processors, faster input–output paths, etc.) once the specific causes have been identified. Similarly, observation of how the operators use the machine can show how particular facilities can be modified or redesigned to make them more convenient to use. Neither of these approaches requires any alteration to the FRONTIERS concept or any change to the division of work between the operator and the machine, but the margin for improvement by these means alone is unlikely to be sufficient for the foreseen increase in work load.

Ultimately, the point must come, though we have not reached it yet, when further enhancements to the computer system will yield negligible overall improvement because the remaining significant delays are due to the thinking time needed by the operator. We must therefore seek other ways of upgrading FRONTIERS so as to give the operator more help, and options (*b*) and (*c*) below offer two ways of doing this.

(b) Interaction with a dynamical weather-forecasting model

The second avenue of development for FRONTIERS takes account of the position of the system as a cooperating part of a much larger forecasting operation. Of particular significance is the current development by the Meteorological Office of a mesoscale numerical forecasting model that will cover the entire U.K. and use a horizontal grid spacing of 15 km (Golding 1987). The model, which solves the equations of motion, state and continuity for the atmosphere, will be run every 6 h and produce forecasts of many parameters, including precipitation, for periods of up to 18 h ahead. Although they cover similar geographical areas and both make short-period predictions, FRONTIERS and the mesoscale model have different strengths and so complement one another rather than compete. The first few hours of any detailed forecast are necessarily dominated by the current situation and recent trends, whereas for longer term forecasts these will be less important than nonlinear development. The mesoscale model predicts these longer term developments, which cannot appear in the FRONTIERS extrapolation forecasts, so that beyond a few hours ahead the numerical model will produce better guidance. However, the model cannot properly represent features with scales of less than three or four grid lengths, so the current situation and its immediate development are shown better by the FRONTIERS products. The point at which the cross-over in skill occurs will vary from case to case, but is likely to be in the range 2–6 h; it will be up to the forecaster who receives both products to judge which advice to follow when they differ.

To allow FRONTIERS and the mesoscale model to operate in isolation would be to throw away important synergetic opportunities for improving both. It is already planned to use FRONTIERS quality-controlled rainfall composites as one source of data for initializing the mesoscale model. Rainfall observations cannot be assimilated directly by the model but instead are used to adjust fields of relative humidity and vertical motion. Experiments will also be done to see whether output from the mesoscale model can be used to provide guidance for FRONTIERS. Among the possibilities are to obtain predictions of development and decay from the model and apply these to the FRONTIERS forecasts, instead of asking the forecaster to define constant development–decay factors as at present, and to use model wind fields to advect the analysed rainfall field rather than extrapolating rainfall clusters with constant velocity as at present. If successful, these methods would allow the automation of a substantial manual process, although the capacity for manual intervention would need to be retained as a safety net.

(c) Application of expert-system techniques

The remaining line of attack is even more challenging. This will be to try to copy, within the computer, some of the judgemental processes which have so far been left entirely to the operator. This is different from the example given in §4*b*, where we propose trying to replace an existing manual method with a different one more suited to automatic computation. Both approaches are likely to find application within FRONTIERS, but this third approach represents a fundamental change in the relation between the operator and the FRONTIERS computer.

FRONTIERS, as implemented at present, is not an 'intelligent' system. The job of the computer is to present control options and image data to the operator and to perform whatever processing functions are necessary to implement his decisions. It is the operator's responsibility to provide all the meteorological knowledge and understanding needed to interpret the data and to make all the decisions, both about modifying individual images and about selecting which functions to perform. The man–machine interface thus acts as a boundary, not only between the man and the machine, but between two quite separate classes of function.

We now propose to try to move some of these functions across the boundary, into the machine. In this context it should be remembered that all the case-specific data used by the operator to guide his decisions reach him via computers, and are therefore, in principle at least, available to FRONTIERS. What are not yet available to the machine, and what we have not begun to formulate in the detail required, are the knowledge and experience which enable the operator to organize and select these disparate, incomplete and sometimes conflicting pieces of observational evidence, and fit them together to form a coherent picture.

The fundamental limit to this transfer of expertise will be set by our ability to understand and formulate the operator's reasoning paths. Until now, the use of the man–machine mix, in which are combined the versatility of the human forecaster and the data-handling power of the computer, has absolved us from the need to formulate the tasks in meticulous detail. Guidance on the use of the FRONTIERS system has been provided, but we have relied on the operator, an experienced forecaster, to be able to take sensible action, even in unforeseen situations.

The availability of a group of operators experienced in using FRONTIERS in its present form, for real analysis and forecasting, will be invaluable in testing and refining our understanding. In addition, FRONTIERS archives all its cases on magnetic tape so that they are available for further analysis; it will thus be possible to test attempts at automation against a large number of real cases and, particularly when they fail, to examine how one or more human operators tackle the same problem.

A valid criticism of interactive systems like FRONTIERS is that the quality of the products depends upon the skill of the individual operator and that operators are not uniformly experienced and skilled. Automation will improve the objectivity of the FRONTIERS processes and make the results less dependent upon the operator, even if he retains a supervisory role with the option of accepting or altering the automatically generated products. It will also allow more use to be made of forecasting rules which depend directly on calculations based on the physics of atmospheric processes. However, it will be important to measure the success of automated functions against the results achieved by human operators; automation is likely to be easier and more successful in some cases than in others and a pragmatic approach to its introduction is needed.

Success in automating FRONTIERS, using either method (*b*) or method (*c*), will shift some of the burden from the operator to the computer, and we must expect to encounter performance problems in the machine as a result. However, these are problems amenable to solution by method (*a*), enhancements to the computing system, rather than being absolute obstacles set by the speed at which a human being can work.

5. Conclusion

The availability of image data from radars and satellites has revolutionized our ability to measure the detailed distribution of precipitation and has led to the emergence of a new class of very short period extrapolation forecasts or 'nowcasts'. New methods have had to be developed to process these extremely perishable data and deliver them to users quickly enough to be of practical value. In the Meteorological Office this has been achieved through the development of a special purpose interactive display system, 'FRONTIERS', which allows us to merge the knowledge and experience of a professional forecaster with the data-handling and presentation capabilities of a computer. The forecaster modifies images and generates forecasts, using information from many sources about the current weather situation to help compensate for the unusual error characteristics of the radar and satellite imagery.

The forecaster is already kept very busy in active weather situations and, with the doubling in size of the radar network during the next few years, he will no longer be able to keep pace with the flow of real-time data. There is some scope for improving throughput by simply increasing the speed of the computer system, but this will be insufficient by itself. Means must be sought of transferring to the computer some judgemental and interpretive tasks, hitherto entirely the province of the human operator. This will be a difficult and challenging task, which will hinge on our ability to make a much more detailed and precise analysis of the forecaster's task than has so far been attempted. In it we shall be greatly helped by the experience gained, using the present FRONTIERS system, of how a forecaster tackles his job, and the ability to test ideas against the large collection of cases archived by FRONTIERS. Success in this enterprise will not be easy to achieve but is likely to have considerable impact, both on FRONTIERS and on many other aspects of the weather-forecasting process.

References

Browning, K. A. 1979 The FRONTIERS plan: a strategy for using radar and satellite imagery for very short-range precipitation forecasting. *Met. Mag., Lond.* **108**, 161–183.

Browning, K. A. & Collier, C. G. 1982 An integrated radar-satellite nowcasting system in the UK. In *Nowcasting* (ed. K. A. Browning), pp. 47–62. London: Academic Press.

Browning, K. A. 1986 Conceptual models of precipitation systems. *Weather forecast.* **1**, 23–41.

Cluley, A. P. & Hills, T. S. 1988 Meteorological Office outstation display system: from concept to reality. *Met. Mag., Lond.* **117**, 1–12.

Golding, B. W. 1987 Short range forecasting over the UK using a mesoscale forecasting system. In *Short and medium range weather prediction* (ed. T. Matsuno). Universal Academy Press Inc.

Lovejoy, S. & Austin, G. L. 1979 The delineation of rain areas from visible and IR satellite data for GATE and mid-latitudes. *Atmos. Ocean* **17**, 77–92.

Suomi, V. E., Fox, R., Limaye, S. S. & Smith, W. L. 1983 McIDAS III: a modern interactive data access and analysis system. *J. Clim. appl. Met.* **22**, 766–778.

Discussion

[*Questioner not identified.*] What is the economic justification for a system such as FRONTIERS?

B. J. Conway. The justification for FRONTIERS is in the value it adds to the weather radar network. The raw radar coverage is incomplete and the data subject to errors; the map of the network that I presented showed the maximum range for *detection* of rain, but the range for reliable quantitative measurements is much smaller and so there are gaps in such coverage. More radars will be added in the next few years but these will mainly serve to extend the outer boundary of the network so as to encompass the rest of the U.K. The cost of FRONTIERS is small compared with that of adding enough radars to fill all the gaps in quantitative coverage, yet it not only helps us to compensate for the deficiencies in the raw data, but it also adds an entirely new dimension to the products by allowing us to produce short-period extrapolation forecasts. The radar network is an expensive investment and we would be neglecting our duty if we failed to extract the maximum value from it; FRONTIERS is our way of getting the best from this facility.

A. M. E. Runacres (*Department of Geography, University College, Swansea, U.K.*). Is it possible for the FRONTIERS operator to alter the depiction of rainfall intensity, as opposed to adding or deleting pixels completely?

B. J. Conway. The forecaster can, within limits, subjectively adjust the overall calibration of any radar; he also has some freedom about how and where he instructs FRONTIERS to calculate and apply quantitative corrections (for example for orographic enhancements) but he cannot adjust the values of individual rainfall pixels just as he chooses.

A. M. E. Runacres. Would Dr Conway comment on the problems encountered in comparing and calibrating the weather radar, which gives instantaneous but areally averaged samples. with raingauges, which give point measurements integrated over a period of time?

B. J. Conway. This causes many difficulties in calibrating and evaluating radar rainfall measurements, particularly in convective situations where the rainfall has high spatial and temporal variability. There is no way of eliminating the problem with any practicable network of gauges. In situations of more uniform, widespread rain the comparison is easier, but even then topographic effects mean that individual gauges are valid as calibration tools only within limited local domains, and these domains vary with factors such as the wind direction. FRONTIERS has access to calibration gauge data but the operator must exercise caution about how they are used.

P. L. C. Jeynes (*Oxford Computer Services Ltd, U.K.*) What wavelengths do the ground-based radars operate at?

B. J. Conway. They operate at 5.6 cm. One or two older radars, now being replaced, operate at 10 cm.

P. L. C. Jeynes. So the radar system essentially relies on measuring the 'Rayleigh' back-scattering from the rain droplets?

B. J. Conway. Yes.

P. L. C. Jeynes. Would there be any useful Doppler information in the echo?

B. J. Conway. Given a sufficiently dense network of Doppler radars, we could obtain useful information about the horizontal divergence field of the air and therefore its vertical motion, but we do not usually have the severe tornadic storms in this country that make Doppler radar the natural choice in the U.S.A.

P. L. C. Jeynes. I was thinking more in terms of using the Doppler shift to measure the vertical free-fall terminal speed from which one might infer the rain droplet size.

B. J. Conway. This would be difficult with current radars which operate near horizontal. The high beam elevations which would be needed for fall-speed measurements would limit the real coverage of such measurements.

P. L. C. Jeynes. Perhaps the vertical rain motion could be obtained from satellite-borne or other radars?

B. J. Conway. Perhaps, though satellite systems have their own problems, such as resolution, power requirements and, for polar orbiters, the interval between observations. With the ground-based radar system we have to choose carefully how we spend our limited resources. Our current priority is to extend the area of coverage to the whole of the U.K.

Phil. Trans. R. Soc. Lond. A **324**, 317–323 (1988)
Printed in Great Britain

The integration of meteorological satellite imagery and numerical dynamical forecast models

By W. L. Smith, L. M. Leslie†, G. R. Diak, B. M. Goodman, C. S. Velden, G. M. Callan, W. Raymond and G. S. Wade

Cooperative Institute for Meteorological Satellite Studies (CIMSS), University of Wisconsin-Madison, Madison, Wisconsin 53706, *U.S.A.*

[Plate 1]

Imagery of clouds, water vapour, and thermal structure as achieved in real-time from geostationary satellites can play an important role in the dynamic verification and reinitialization of numerical weather-prediction models used for short-range prediction of intense weather. By overlaying hourly interval model forecast fields over satellite imagery, the validity of the evolving forecast can be subjectively assessed and phase and amplitude errors can be diagnosed in a dynamic manner. Videographic interactive computer systems designed for the purpose should enable improved numerical weather forecasts to be made through the dynamic use of satellite imagery. This approach is demonstrated by using the Man–Computer Interactive Data Access System (McIDAS) at the University of Wisconsin for a case of the development of a cyclone over the east coast of the U.S.A.

1. Introduction

For future numerical prediction of localized weather, satellite imagery will play a key role in two areas: (*a*) the provision of quantitative wind estimates from tracking clouds and inhomogeneities in water vapour by using animated multispectral imagery from geostationary satellites, and (*b*) the diagnosis of model performance from quasi-continuous satellite imagery. For the latter application, the departure of the time continuity of a model variable (e.g. vertical motion) from the time continuity of cloud, water vapour, and/or thermal features displayed in imagery provided from the geostationary satellite are assumed to be indicative of the model forecast deviating from reality. For both applications, the use of a videographic interactive computer system plays a key role. It enables manual interaction for tracking cloud and water-vapour elements as displayed in an animated sequence of images on a video monitor, and for the graphic overlay of contour analyses of the output from a numerical weather prediction (NWP) model at the time of the animated satellite imagery.

The introduction of videographic interactive computer systems into forecast offices during the next several years will enable the marriage of NWP and satellite imagery for mesometerological applications. The prefix 'meso' denotes the scale of relatively small and short-lived weather systems such as fronts, convective storms and sea breezes, in contrast to the term 'synoptic', which denotes the scale of relatively large and long-lived weather systems such as major travelling low- and high-pressure systems and monsoons. In such a setting, the forecaster can view at frequent time increments (e.g. hourly) the performance of the model and use this either to correct subjectively the numerical prognosis for the future or to have the model

† Permanent address: Bureau of Meterology Research Centre, Melbourne, Australia.

reinitialized with current data and thereby generate an updated numerical forecast sequence. An example of the former is where the forecaster observes from a time sequence of model output (e.g. predicted surface pressure) and satellite imagery (e.g. visible cloud cover) that the model is moving a frontal system too quickly or slowly compared with nature. If the developing phase error appears to be linear in time, the forecaster could subjectively alter the validation time of subsequent numerical prognosis so as to conform more properly to reality. An example of the latter situation is where the forecaster sees a weather disturbance begin to develop in the satellite imagery that is not evolving in the sequence of numerical prognoses. In this case, the course of action is to analyse the current observations which would contain information on the new disturbance and then to reinitialize the forecast sequence.

The purpose of this paper is to discuss briefly the elements of such an experimental interactive numerical forecast capability being developed at the University of Wisconsin. An example is presented of the application of this capability to a case of cyclogenesis along the east coast of the United States observed during the second intensive observation period (IOP) of the Genesis of Atlantic Lows Experiment (GALE). The example demonstrates that systems that use videographic interactive computer technology for mesoscale NWP model inititalization and real-time diagnosis of model performance can and should play a key role in the mesoscale weather-analysis–forecast operation of the next decade.

2. The University of Wisconsin videographic computer system: McIDAS

The Man–Computer Interactive Data Access System (McIDAS) has evolved at the University of Wisconsin (Suomi *et al.* 1983) into a powerful tool for (*a*) processing and amalgamating meteorological data from *in situ* and remote sensors, (*b*) displaying contour analyses of these meteorological data over imagery from satellite, radar and lightning sensors for nowcasting applications, and (*c*) preparing and disseminating forecasts of future weather events based upon the timely and comprehensive analyses of the evolving weather patterns. Recently, a numerical weather prediction (NWP) component has been added to McIDAS. The primary purpose of the NWP component, as described briefly in the next section, is to enable four-dimensional assimilation of conventional and satellite derived moisture, wind and mass-field observations. This process will enable the production of mesoscale analyses of the atmospheric state variables and implied vertical motion, precipitation, cloud cover and other weather variables on a quasi-continuous basis. A secondary purpose is to conduct extended time numerical forecasts of mesoscale weather to demonstrate the use and potential impact of new ground-based and satellite remotely sensed wind, temperature and moisture data. As the McIDAS technology becomes an integral part of operational weather forecast offices, methods must be developed to optimize the combined use of the interactive videographic display capability, remotely sensed mesometeorological data and numerical analyses and forecasts of the atmospheric state variables.

The McIDAS also plays a key role in the generation of wind, mass and moisture observations remotely sensed from meteorological satellites (Smith *et al.* 1981*a*). For example, cloud and water-vapour winds are generated by a meteorologist viewing an animated sequence of geostationary satellite images. The operator manually selects features that are temporally consistent and, therefore, appear to be reliable tracers of atmospheric motion (see, for example, Hayden & Stewart 1987). Computer software enables accurate determination of the wind vector once the appropriate target has been selected. Quantitative interpretation of the infrared signal enables a suitable altitude assignment of the vector to be made. Figure 1, plate 1, shows an example of the cloud and water-vapour wind data for one of the initial analysis times of the case study presented in this paper.

Vertical-temperature and water-vapour soundings from both polar orbiting and geostationary satellites are achieved at the University of Wisconsin by using McIDAS to enable interactive manual control in the data editing and enhancement of the data density in meteorologically important regions (Smith *et al.* 1979, 1981*b*). The interactive capability for producing soundings is important because a skilled meteorologist can differentiate significant fine-scale meteorological phenomena from small-scale features produced by errors in the remotely sensed data.

3. The CIMSS numerical weather-prediction model

The numerical weather prediction model component of McIDAS is a fifteen-level high-resolution (30–60 km) primitive equation model originally written for the Australian region (McGregor *et al.* 1978) and adapted for use in the North American region. The model has since been amended through the use of more efficient numerics, a more complete boundary layer parametrization (Leslie *et al.* 1985), the inclusion of improved numerical procedures to handle better the high topographic regions of the U.S. (Corby *et al.* 1972), and the replacement of the original parametrization for predicting convective rainfall with an improved scheme (Hammarstrand 1977).

The three-dimensional objective analysis component of the model is an evolution of the scheme developed in the Australian Numerical Meteorology Research Centre (ANMRC) by R. S. Seaman (1977) for limited area analysis over the Australian region, and adapted to the U.S. region (Mills & Hayden 1983). The model describes the atmosphere in terms of the geopotentials of constant pressure surfaces. The geopotential is the height of a surface normalized with respect to the local gravitational acceleration. The geopotential 'thickness' of a layer bounded by two pressure surfaces is proportional to the mean temperature of the air within the layer. The analysis system is based upon the calculus of variations used in an explicitly three-dimensional manner. The basic observational ingredients combined in the analysis are observed geopotential thicknesses (derived from the atmospheric temperature) of each layer and horizontal geopotential gradients either from geopotential measurements or independently derived from wind measurements and a wind law that relates the horizontal gradients of geopotential to the observed winds. Reliability weights for the analysed geopotential and geopotential gradient fields are determined on the basis of data density and standard deviation parameters which vary with latitude and level. In this way, the blended geopotential fields reflect a greater weight to the wind data in low latitudes and to the temperature information at higher latitudes.

In the forecast application of the model, the analyses are performed by using conventional and any special observations of surface and upper air pressure, temperature, moisture and wind (e.g. the special GALE observations used in the example presented in the next section), satellite-observed wind and vertical-temperature and moisture-profile data, as well as sea-surface temperature derived from the vertical atmospheric sounder (VAS) on the U.S. geostationary satellite (Bates *et al.* 1988), which is useful for establishing surface layer heat fluxes (Diak *et al.*

1986). Because of the large-scale systematic differences often existing between the geopotential heights derived from VAS and conventional data, the VAS data are used in horizontal gradient form in the variational analysis (Diak 1988). Examples of the data coverage for the GALE observation area are shown in figure 2.

For the forecast exercise, the model is nested within the global forecast model of the National Meteorological Center (NMC) of the U.S.A.; that is, the NMC global forecast is used to provide the values of the atmospheric variables at and across the lateral boundaries of the region shown in figure 2. The boundary conditions at each model time step are produced by linear time interpolation of successive NMC 12 h global forecast fields.

4. An example of the use of imagery in numerical weather prediction

A demonstration of the use of satellite imagery in diagnosing the performance of a NWP model is provided here for the case of the rapid development of a cyclone off the east coast of the United States on 27 January 1986. For this case, the model was initialized at 00h00 GMT on 27 January with data obtained by the conventional surface and upper-air network and special radiosonde, dropsonde, ship, buoy, and satellite cloud and water-vapour wind data provided as part of the program (figure 2). The initial mean sea-level pressure field (MSLP) is shown in figure 3*a*, plate 1. As can be seen, at this time, there is a synoptic-scale low-pressure system centred off the Carolina coast with an intense cold front visible in the cloud imagery trailing southward across the western Atlantic and into southern Florida. During the next 24 h, the low-pressure centre intensified and moved northward across the northeastern United States and into southeastern Canada with the cold front propagating well out into the Atlantic Ocean (figure 3*e*). Inspection of the hourly interval output of the 24 h forecast sequence with respect to the cloud and water-vapour imagery from *GOES* at hourly time intervals revealed that the model successfully captured the intensification and movement of the synoptic-scale low-pressure system and the propagation of its associated cold front (figure 3*d*). However, the model failed to forecast a secondary development behind the front at about 18 h into the forecast cycle (see figure 3*b*,*c*). The new disturbance can be clearly seen in the *GOES* cloud imagery at 18h00 GMT (figure 4, plate 1).

Description of Plate 1

FIGURE 1. Satellite-derived winds from tracking inhomogeneities in clouds and water vapour for 00h00 GMT 28 January 1986. The water-vapour winds are shown overlaying 6.7 μm water-vapour imagery from the *GOES* satellite, whereas the cloud-drift winds are depicted overlaying 11.0 μm cloud imagery.

FIGURE 3. Mean sea level pressure (MSLP) analyses and forecasts for 00h00 GMT and 18h00 GMT 27 January 1986 and 00h00 GMT 28 January 1986. (*a*) Analysis for 00h00 GMT, (*b*) forecast for 18h00 GMT, (*c*) analysis for 18h00 GMT, (*d*) 24 h forecast for 00h00 GMT, (*e*) analysis for 00h00 GMT, and (*f*) 6 h forecast for 00h00 GMT.

FIGURE 4. Composite visible infrared *GOES* imagery at 18h00 GMT on 27 January 1986. The imagery reveals the development of a storm system over the Atlantic Ocean just off the Virginia–Maryland coast of the United States.

FIGURE 5. Original 24 h forecast and subsequent reinitialized 6 h forecast of vertical motion ($\mathrm{d}p/\mathrm{d}t$) at the 7×10^4 Pa pressure level for 00h00 GMT 28 January 1986. Negative values refer to upward motion, whereas positive values denote downward motion. The isolines of vertical motion overlay the satellite cloud imagery at the forecast validation time. The imagery has been enhanced such that the darkest regions within the otherwise white cloud mass correspond to the highest and brightest cloud cells. The 6 h forecast has benefitted from the reinitialization as explained in the text.

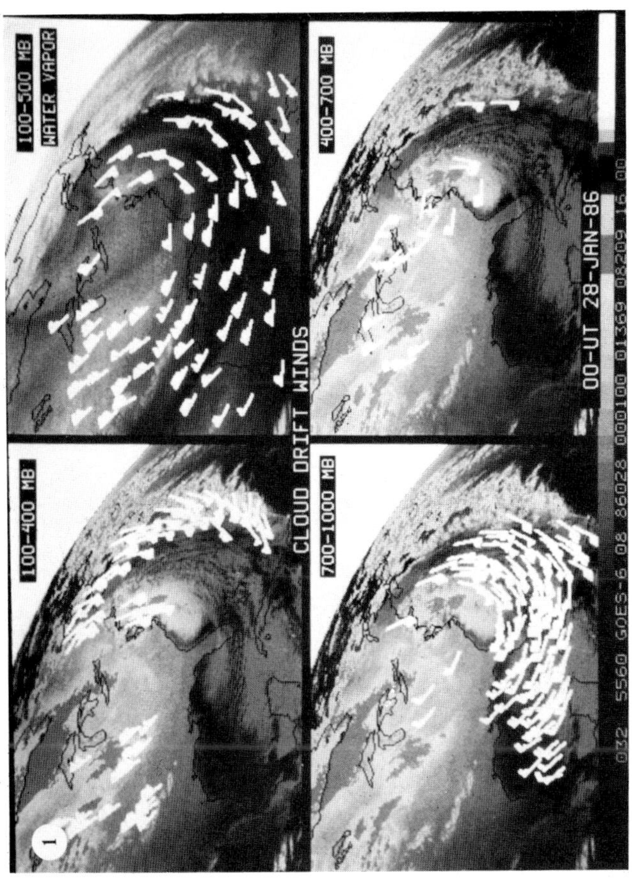

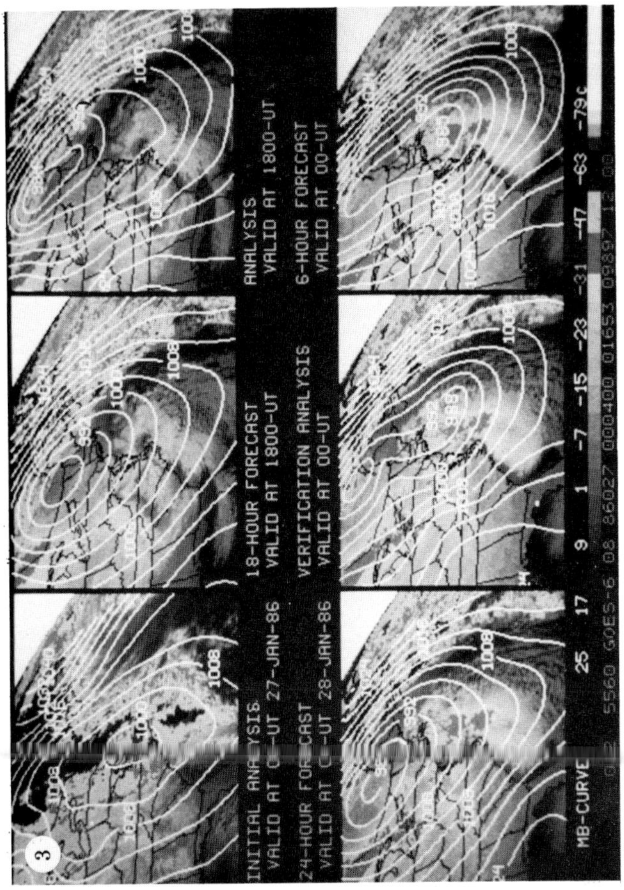

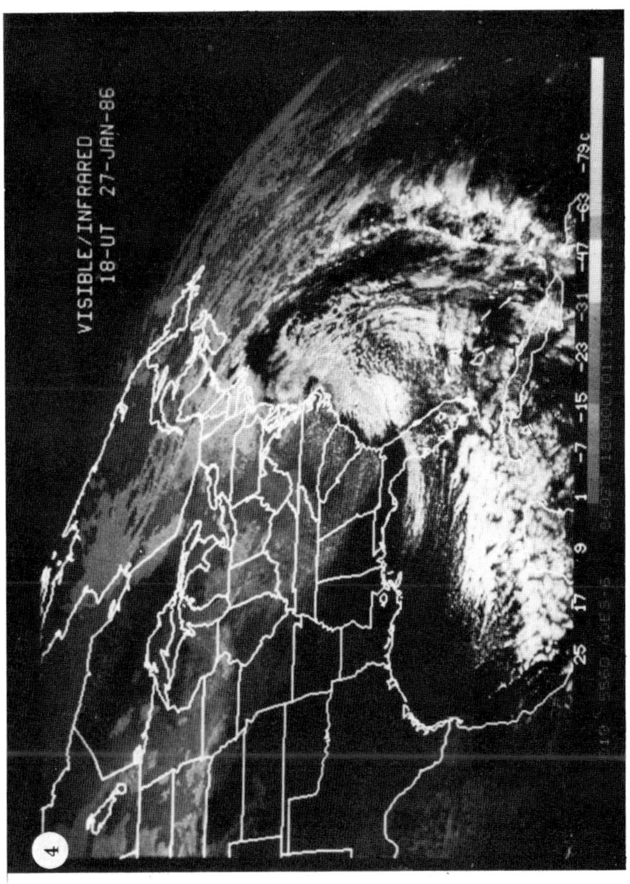

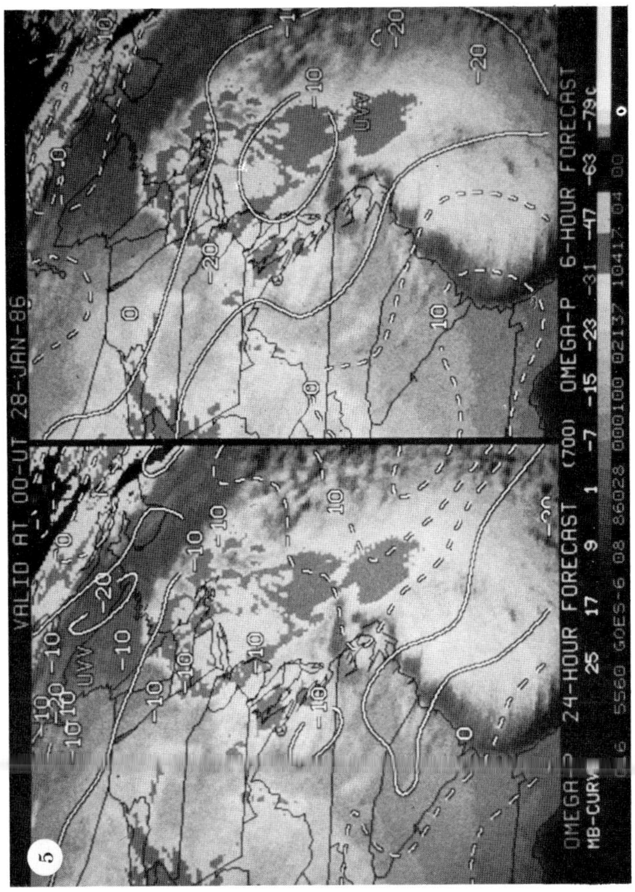

PLATE 1. For description see opposite.

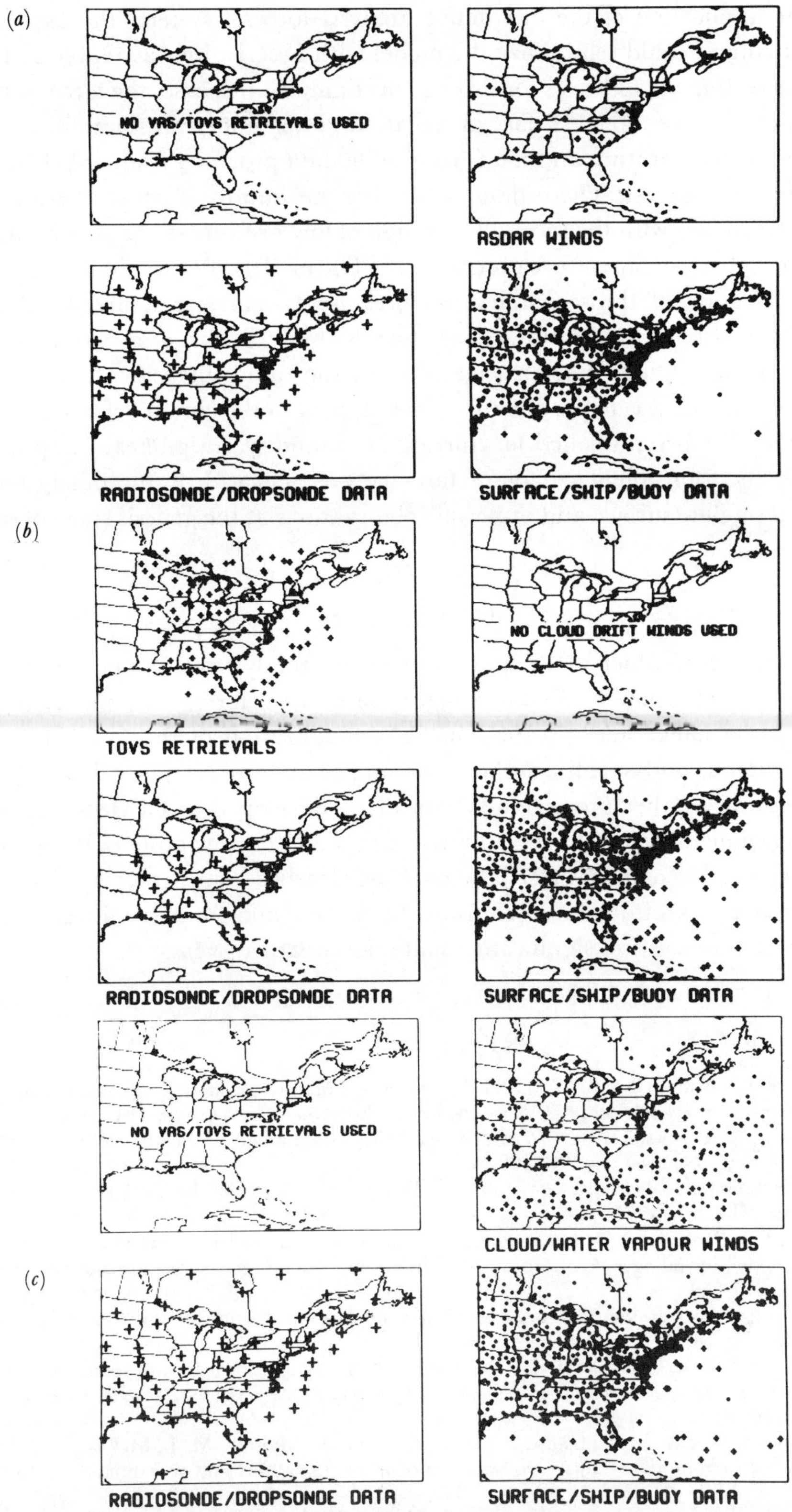

FIGURE 2. Coverage of data used in numerical analyses for (*a*) 00h00 GMT 27 January 1986 (upper four panels), (*b*) 18h00 GMT 27 January 1986 (Middle four panels) and (*c*) 00h00 GMT 28 January 1986 (lower four panels). VAS (VISSR atmospheric sounder) and TOVS (Tiros operational vertical sounder) retrievals refer to the soundings observed from the U.S. geostationary and polar orbiting satellites, respectively. ASDAR (aircraft satellite data acquisition relay) winds denote the automated wind reports received from commercial aircraft.

In a real-time application of the interactive analysis–forecast system, the user, seeing the apparent model failure, could reinitialize the model with observations available at 18h00 GMT in the hope of correcting the forecasts for subsequent times. In this case, the forecast model was reinitialized with the conventional surface observations, GALE upper-air soundings, and a large volume of upper-air temperature and moisture-profile data provided from a *NOAA*-9 satellite pass over the area of interest. These data resulted in an improved atmospheric analysis at 18h00 GMT (see figure 3*c*) with the secondary trough of low pressure diagnosed by the surface-pressure and atmospheric-temperature observations. The model, after reinitialization, forecasts the rapid intensification of the secondary low-pressure system (see figure 3*c*,*f*,*e*). The re-initialized model also forecast the upward motion within the lower troposphere along the northeast coast of the United States associated with the rapid intensification (see figure 5, plate 1). This disturbance was to produce 7–11 cm of snow in the New York area, whereas the official weather service forecast called for flurries. In summary, a significant improvement in the numerical forecast for 00h00 GMT of 28 January was achieved by the timely inclusion of conventional and satellite surface and upper-air observations at the genesis stage of small-scale cyclone development.

5. Conclusion

Mesoscale numerical weather prediction will benefit greatly from the use of videographic computer technology. It enables satellite imagery to be used for (*a*) the timely diagnosis of model forecast performance and (*b*) subsequent reinitialization of the forecast model with available conventional, and satellite-surface and upper-air observations. The experimental interactive numerical weather prediction system being developed at CIMSS is to serve as a prototype of operational systems for the 1990s and beyond. The example of its application to the east coast cyclone development of 27 January 1986 clearly demonstrates that such a system, utilizing a dense network of observations from the surface and from satellites, is capable of producing improved forecasts of significant small-scale weather events.

References

Bates, J. J., Smith, W. L., Wade, G. S. & Woolf, H. M. 1988 An interactive method for processing and display of sea surface temperature fields using VAS multispectral data. *Bull. Am. met. Soc.* (In the press.)

Corby, G. A., Gilchrist, A. & Newson, R. L. 1972 A general circulation model of the atmosphere suitable for long period integrations. *Q. Jl R. met. Soc.* **98**, 809–832.

Diak, G. 1988 Assimilation of scalar versus horizontal gradient information from the VAS into a mesoscale model. *J. Clim. appl. Met.* (In the press.)

Diak, G., Heikkinen, S. & Bates, J. 1986 The influence of variations in surface treatment on 24-hour forecasts with a limited area model, including a comparison of modelled and satellite-measured surface temperature. *Mon. Weath. Rev.* **114**, 215–232.

Hammarstrand, U. 1977 On parametrization of convection for large scale numerical forecasts at mid-latitudes. *Contr. Atmos. Phys.* **50**, 78–88.

Hayden, C. M. & Stewart, T. R. 1987 An update on cloud and water vapour tracers for providing wind estimates. In *6th Symposium on Meteorological Observations and Instrumentation, New Orleans, 12–16 January*, pp. 70–75. American Meteorological Society.

Leslie, L. M., Mills, G. A., Logan, L. W., Gauntlett, D. J., Kelly, G. A., Manton, M. J., McGregor, J. L. & Sardie, J. M. 1985 A high resolution primitive equations model for operations and research. *Aust. Met. Mag.* **33**, 11–35.

McGregor, J. L., Leslie, L. M. & Gauntlett, D. J. 1978 The ANMRC limited-area model: consolidated formulation and operational results. *Mon. Weath. Rev.* **106**, 427–438.

Mills, G. A. & Hayden, C. M. 1983 The use of high resolution satellite temperature and moisture profiles to intialize a mesoscale numerical weather prediction model – a severe weather event case study. *Mon. Weath. Rev.* **22**, 650–663.

Seaman, R. S., Falconer, R. L. & Brown, J. 1977 Application of a variational blending technique to numerical analysis in the Australian region. *Aust. Met. Mag.* **25**, 3–23.

Smith, W. L., Woolf, H. M., Hayden, C. M., Wark, D. Q. & McMillin, L. M. 1979 The TIROS-N operational vertical sounder. *Bull. Am. met. Soc.* **60**, 1177–1187.

Smith, W. L., Nagle, F. W., Hayden, C. M. & Woolf, H. M. 1981*a* Vertical mass and moisture structure from TIROS-N. *Bull. Am. met. Soc.* **62**, 388–393.

Smith, W. L., Suomi, V. E., Menzel, W. P., Woolf, H. M., Sromovsky, L. A., Revercomb, H. E., Hayden, C. M., Erickson, D. N. & Mosher, F. R. 1981*b* First sounding results from VAS-D. *Bull. Am. met. Soc.* **62**. 232–236.

Suomi, V. E., Fox, R., Limaye, S. S. & Smith, W. L. 1983 McIDAS III: a modern interactive data access and analysis system. *J. Clim. appl. Met.* **22**, 766–778.

Phil. Trans. R. Soc. Lond. A **324**, 325–346 (1987)
Printed in Great Britain

Satellite image analysis techniques applied to oceanography

By P. E. La Violette

Naval Ocean Research and Development Activity, NSTL, Bay St. Louis, Mississippi 39529-5004, *U.S.A.*

[Plates 1 and 2]

Infrared and visible spectral sensors on spacecraft can provide imagery of ocean conditions equivalent to a global network of *in situ* sensors. Oceanographers can now view all the world's oceans during one day's travel of a polar-orbiting satellite. The surface areas of complex ocean regions that have strong thermal and chemical gradients, such as the Gulf Stream in the Atlantic and the Kuroshio in the Pacific, are displayed in minute detail in the imagery provided by satellite infrared and visible sensors.

Because of the vast quantities of data involved in satellite infrared and visible imagery, computer techniques must be used to derive oceanographic information. A variety of basic computer image analysis techniques are required to do so. These basic techniques include selective enhancement, geographic registration, absolute ocean radiation and multiple-image composition. Through the use of these techniques, satellite and conventional oceanographic data can be integrated to form synergistically a powerful analytical tool for modern ocean research.

Introduction: general concepts

Although seemingly uniform to the uninitiated, the oceans of the world are highly variable; any given region has unique distinguishing physical characteristics. These characteristics are controlled by specific factors: the atmosphere, the seasonal migration of the Sun, the restriction of bottom topography, and the influx of thermally–chemically different waters from outside the region. Because these factors vary globally, the resulting characteristics of the ocean vary from region to region.

Many variations can be detected in infrared and visible satellite imagery. How this detection can be done with computer image analysis techniques is demonstrated in figures 1 and 2, plates 1 and 2. But before discussing these techniques, certain basic concepts must be understood.

The first and most important concept involves understanding the physics and the limitations of infrared and visible satellite data. For example, because of their altitude, satellites obviously have no direct contact with the ocean. Their instruments are restricted to sensors that detect the ocean's reflected or emitted electromagnetic radiation (figure 3). Since the atmosphere lies between the ocean and the satellite, the data received by the satellite are actually a summation of both the atmospheric and the oceanic signals. The emanations of the two régimes must be separated to derive quantitative oceanographic data. This separation, which will be demonstrated, can be difficult and requires the application of a second concept.

The second concept (and the main purpose of this paper) recognizes that satellite imagery produces vast amounts of data and that computer analysis techniques are required to

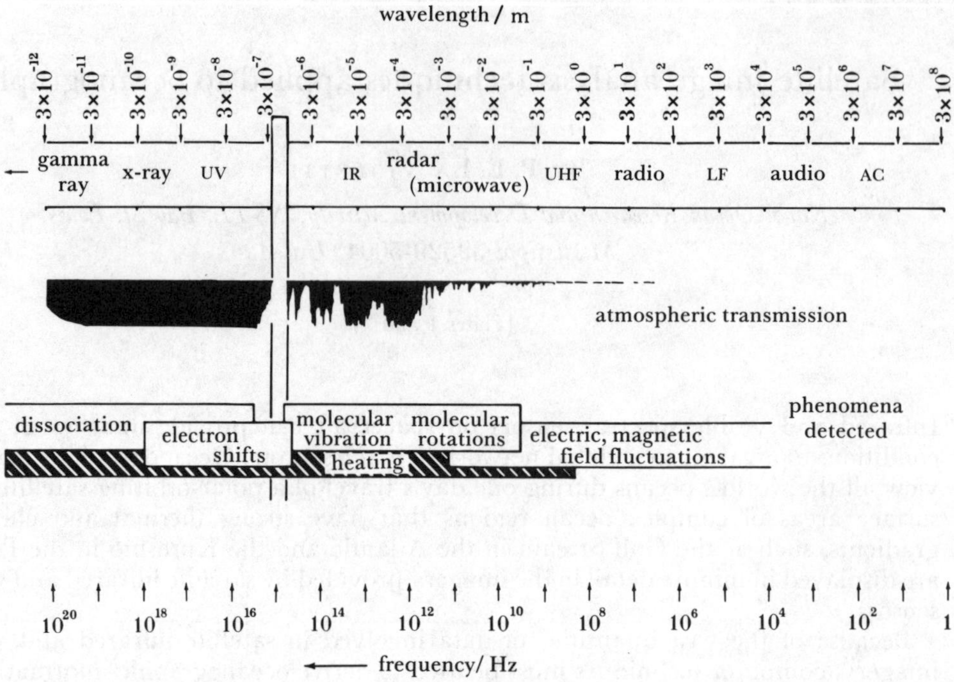

FIGURE 3. Characteristics of the electromagnetic spectrum significant to remote sensing of the ocean.

manipulate these data into usable form. Such techniques can be applied to satellite data by using small computers that interface with image processing and display hardware. These systems are available commercially and are relatively inexpensive. They are a prerequisite for any institution wishing to include satellite image processing as part of its oceanographic programme.

Various computer image analysis techniques can be applied to the study of a particular ocean region. However, in almost all selections, a basic set of techniques should be included: selective enhancement, geographic registration, absolute ocean radiation, and multiple-image composition. These basic techniques will be stressed here. As our skills and knowledge increase, these techniques will improve and, undoubtedly, new techniques will develop.

A final and very important concept is that the satellite data alone are not a panacea for studying the oceans. Satellite data are useful in that alone or in combination with other data

---

DESCRIPTION OF PLATE 1

FIGURE 1. (*a*) This cloud-free *NOAA*-7 AVHRR IR (channel 4, i.e. 10.3–11.3 μm) image of the eastern seaboard of the United States was taken on 28 April 1983. The registered, calibrated image shows the thermal convolutions of the Gulf Stream as it forms a border between the warmer waters of the Sargasso Sea and the coastal cold slope waters off the continental shelf (the darker portions of the picture are the warmer radiated temperatures of the scene).

The image shows the development of Gulf Stream meanders and cold- and warm-core eddies (rings) that were generated when the meanders pinched off. The warm-core ring at 67 °W is easily distinguished as being warmer than the older one northeast of Norfolk, Virginia. This westernmost ring entrained cooler slope water during its journey to this position and now has a cooler centre at the surface with warmer water on the edges. Other features are readily evident. Note the dark regions in the southern area that developed as a result of daytime heating. (Photograph courtesy of Jeffrey Hawkins, NORDA.) (*b*) Note that the details in the cold-water area northwest of the Gulf Stream in figure 1*a* are difficult to discern. In this enhancement the grey tones are set to display that region with the result that the area southeast of the Gulf Stream is poorly defined.

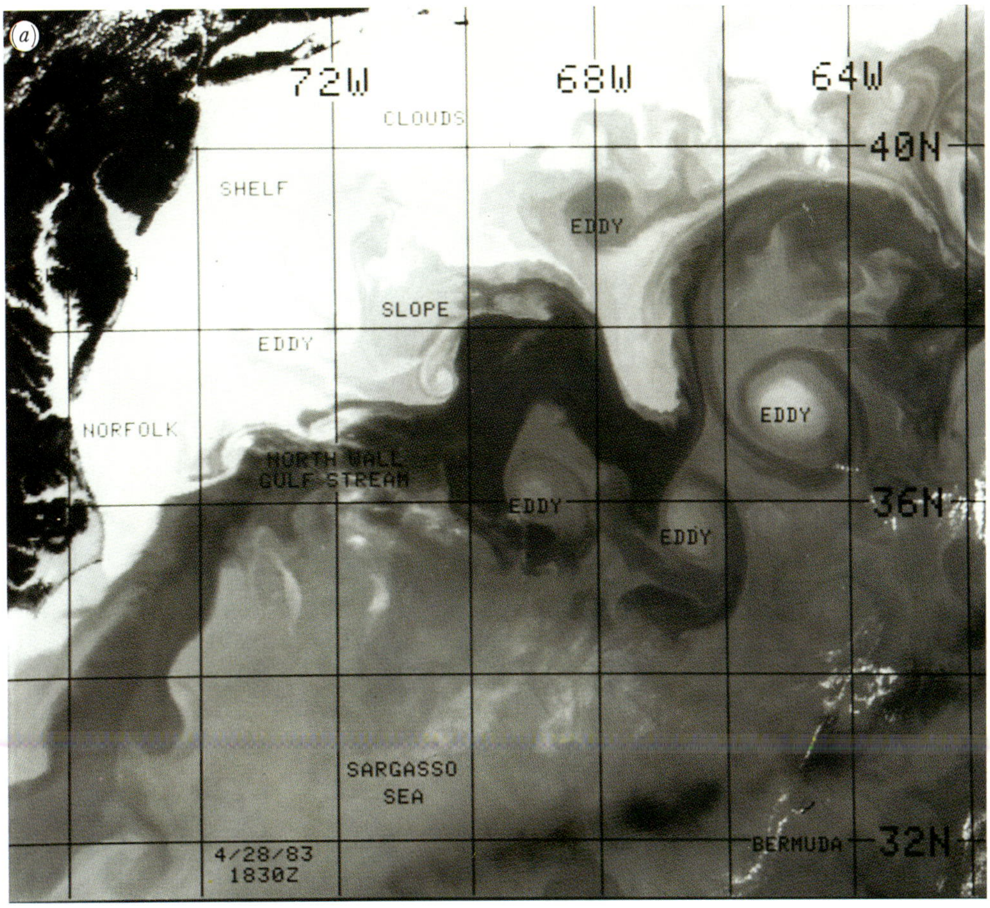

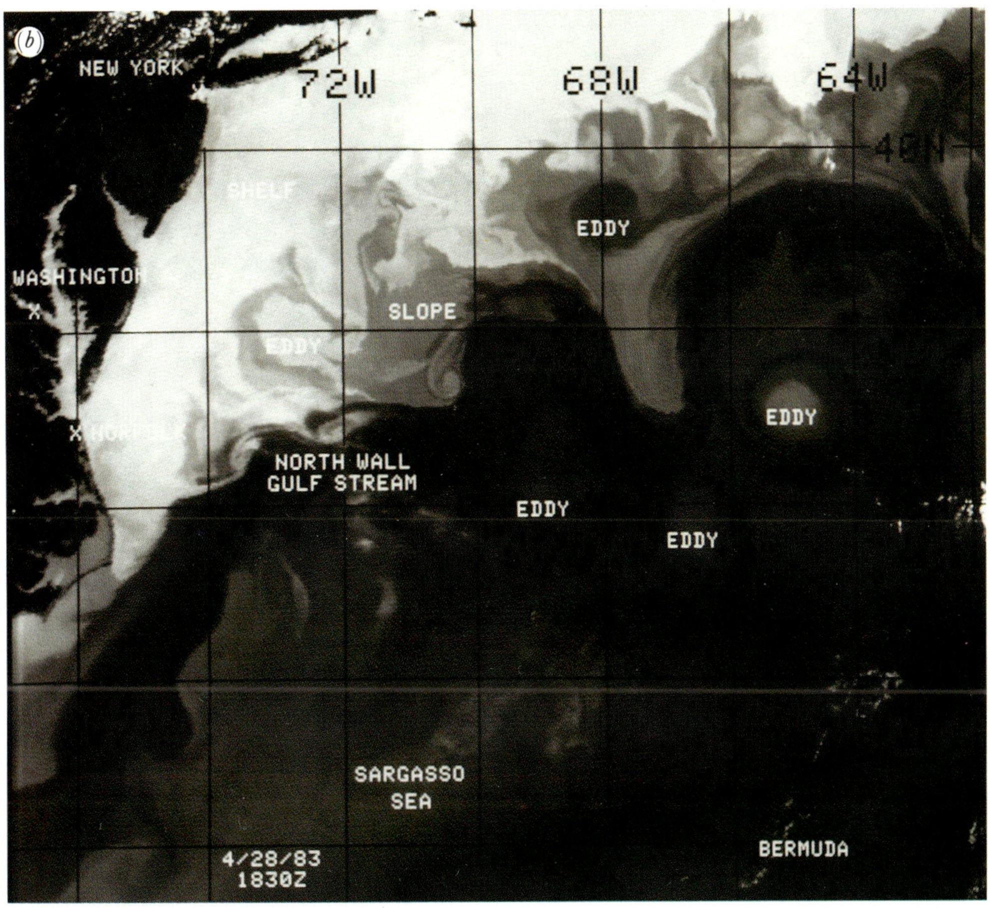

FIGURE 1*a*, *b*. For description see opposite.

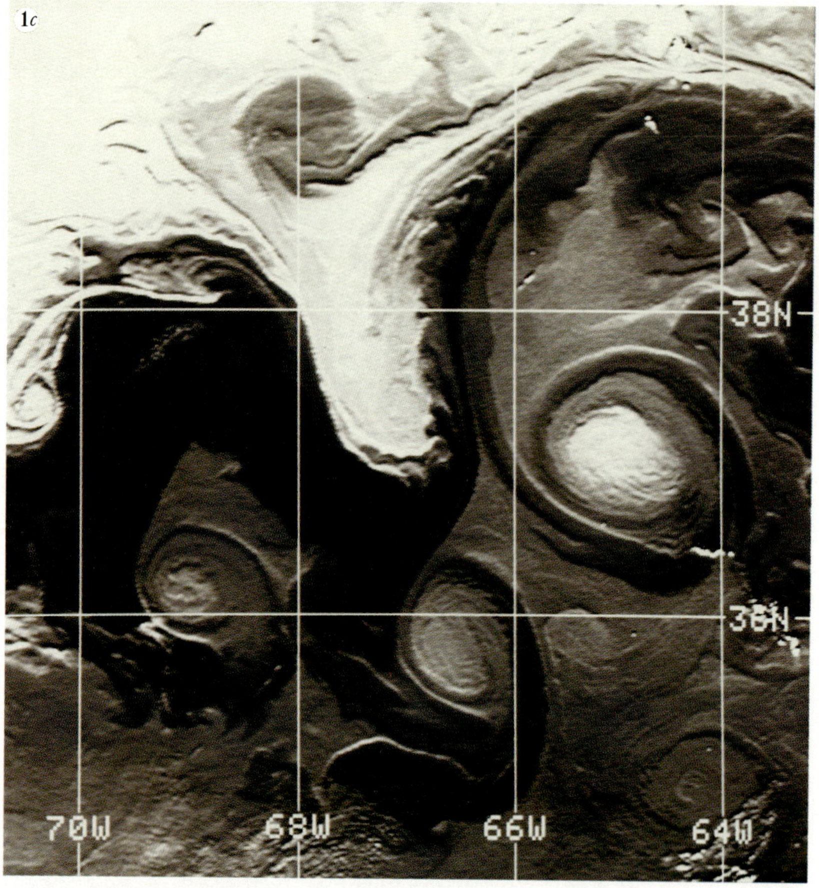

FIGURE 1*c*. For description see opposite plate 1.

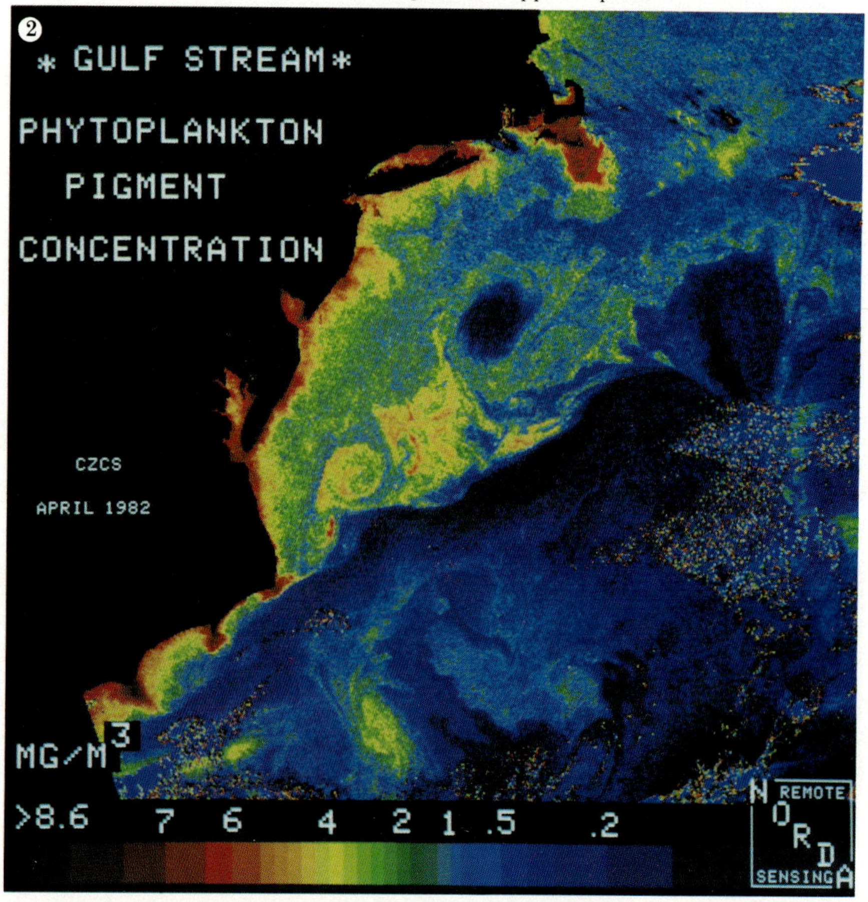

FIGURE 2. For description see opposite.

bases their main output is information. The intuitive skill of the ocean investigator will determine how well the data provide that information. In using satellite data, one must ask certain relevant questions. What created the surface effects seen in the imagery? How deep do these surface effects go? How do these surface effects relate to the characteristics of the deeper water? What conditions are normal? What conditions are abnormal?

In keeping with these concepts, the following sections discuss the physics and the limitations of infrared and visible satellite data, the computer analysis techniques needed to exploit these data, and examples of combined satellite and conventional oceanographic data analyses of two ocean regions.

Infrared and visible satellite data

Infrared satellite data

Infrared satellite data can be used to provide the radiometric surface temperature of the ocean. Ignoring atmospheric effects for the moment, observations in selected spectral ranges can be used to measure the equivalent blackbody temperatures of the ocean surface. In ocean regions where wind speeds are greater than 10 kn†, these temperatures are closely related to what oceanographers consider to be ocean surface temperature (i.e. the temperature of the approximate first metre of water).

Obtaining the oceans' surface temperatures from the altitude of a satellite presents a number of problems, however, and an understanding of these problems is necessary before the data can be used in ocean studies. The following notations are adapted from Holter (1970).

Planck's fundamental radiation equation gives the spectral distribution of radiation from a self-emitting perfect radiator (blackbody) with uniform temperature ($W_{B\lambda}$). Planck's law applies to the entire electromagnetic spectrum and can be expressed in terms of spectral radiant emittance as

$$W_{B\lambda} = C_1 \lambda^{-5} [\exp (C_2/\lambda T) - 1]^{-1}, \qquad (1)$$

where C_1 and C_2 are radiation constants, λ is the wavelength, and T is the temperature of the radiating surface. The units of the equation are 10^{-7} J cm^{-2} s^{-1} μm^{-1}.

† 1 kn = 1.852 km h^{-1} ≈ 0.514

Description of plate 2

Figure 1. (*c*) This higher resolution (1 km) subsection of figure 1*a* was processed with a modified Frei-Chen edge enhancement filter. Filter weights were chosen so that the intensity at each point was replaced by a 3 × 3 finite difference approximation to the original image. Linear contrast enhancement was applied to the image before the filtering. The results show the impression of relief based on the sea surface temperatures. The relief is the opposite of reality in that the white (cold-core) rings should represent a depression in the ocean surface and not a rise. (Photograph courtesy of Matthew Lybanon, NORDA.)

Figure 2. This *Nimbus* 7 czcs false-colour image of the same ocean area as figure 1 was taken on 24 April 1982. In this instance, the registered, calibrated image has been atmospherically corrected and presents colour variations representing values that result from ratioing the czcs 450 and 550 spectral channels. Although the values derived from this ratio are normally used to show the chlorophyll distribution in the ocean, here they are equally instructive in showing the region's current and frontal distribution. Note that the temperature values presented as grey tones in figures 1*a*, *b* could also have been displayed in colour in the way shown here. The advantage of colour goes beyond aesthetics in that the colour scale can show a greater range than the black and white scale. A case in point is that the details of the ocean features northwest of the Gulf Stream are clearly shown in this figure, whereas they had to be added as a subimage in figure 1. (Photograph courtesy of Robert A. Arnone, NORDA.)

In Nature, of course, most objects are not perfect radiators. The radiative efficiency factor, ϵ, or spectral emissivity, of these imperfect radiators (greybodies) varies with wavelength and may be defined as

$$\epsilon_\lambda = W_\lambda / W_{B\lambda}, \qquad (2)$$

where W_λ is the spectral radiant emittance of a greybody for a given wavelength. The spectral emissivity of different objects may range from close to zero for poor radiators to unity for perfect radiators.

Such greybodies as the ocean approach being blackbodies within selected spectral bands. The important issue in most cases, therefore, is the characteristic emissivity for the specific band being monitored. Equations (1) and (2) can be used to obtain the spectral radiant emittance of any surface of known temperature and spectral emissivity.

The surfaces of greybodies reflect, as well as emit energy. Kirchhoff's law states that if a surface is optically opaque, then the spectral reflectivity, ρ_λ, and the spectral emissivity, ϵ_λ, of the surface have the relation

$$\rho_\lambda = 1 - \epsilon_\lambda. \qquad (3)$$

From this relation we know that infrared radiation from a surface is the result of a combination of self-emission of the surface and the reflection of a portion of the incident radiation on that surface.

If we apply these principles to the oceans, the infrared radiation emitting from the ocean surface comes from (*a*) the self-emission of the topmost radiating molecule of water and (*b*) the reflection of portions of energy incident from such atmospheric sources as clouds, water vapour, carbon dioxide and aerosols. The sum of these two radiation components can be defined as the effective spectral radiance emittance of the ocean surface, $W_{e\lambda}$:

$$W_{e\lambda} = \epsilon_\lambda W_{B\lambda} + \rho_\lambda W_{a\lambda}, \qquad (4)$$

where the self-emission component is $\epsilon_\lambda W_{B\lambda}$ and the surface reflection component is $\rho_\lambda W_{a\lambda}$.

Where the effective spectral radiance of the sea surface is measured from satellite altitudes, the additional effect of infrared absorption and emission by atmospheric constituents (chiefly, gas molecules of water vapour, carbon dioxide, nitrogen and ozone) must be considered. At these altitudes, N, the upward spectral radiance at a given zenith angle, θ, and wavelength, λ, is given by

$$N_\lambda(\theta) = W_{e\lambda}(T_s)\,\tau_{\lambda a}(\theta) + \int_{\text{(ocean)}}^{\text{(atmosphere)}} W_{a\lambda}(T_a)\,d\tau_\lambda(\theta), \qquad (5)$$

where $W_{e\lambda}$ is the effective spectral radiant emittance at temperature T_s, $\tau_{\lambda a}$ is the spectral transmissivity through the atmosphere, τ_λ is the spectral transmissivity through a layer that extends from a given altitude to the top of the atmosphere, and $W_{a\lambda}$ is the spectral radiant emittance of the atmosphere at temperature T_a.

The first term on the right of (5) is the radiant emittance of the ocean surface to space; the second term is the emittance of a cloud-free atmosphere, integrated from the ocean surface to the top of the atmosphere.

Molecular absorption by atmospheric gases is important at infrared wavelengths. As mentioned, the important absorbing gases in the infrared are water vapour, carbon dioxide, nitrogen and ozone. In those spectral ranges where one or more of these gases are strongly absorbent, the atmosphere for all practical purposes is opaque. Conversely, in some regions, the absorption by these gases is weak.

It is these latter regions, or atmospheric 'windows', that must be used to measure the ocean's effective spectral radiation (figure 3). The boundaries of these windows are difficult to define because the transmissivity of the atmosphere depends on the concentrations of the active absorbing gases present (thus, the distance the radiation must travel through the atmosphere becomes a factor, because a low-angle look at an ocean region would pass through a thicker atmosphere and, hence, larger amounts of absorbing gases).

The amount of energy radiated by a blackbody surface is given by Planck's function. From (1) and figure 3, it can be seen that the 8–12.5 μm window regions have the maximum emitted energy at temperatures near 300 K. This range, then, is the most effective for oceanographic use and the one most used by passive ocean thermal sensors. In the 3.4–4.2 μm window, reflected sunlight and emitted radiation are almost equal in terms of energy. Thus, during daylight hours, data in this spectral range are contaminated by reflected sunlight.

Visible satellite data

In the visible portion of the spectrum (0.4–0.8 μm), the main signal-generating mechanism from the ocean is the backscatter of incident sunlight (the other source is comparatively minor: the fluorescence of marine microorganisms). Visible remote sensing is limited to day periods and regions where the Sun is at least 30° above the horizon. The ocean's ambient light is reflected into space in many ways, some of which include Fresnel reflection from the surface, diffuse scattering by surface foam, and scattering and absorption within the water column.

Fresnel reflectance tells us nothing directly about the biological–chemical constituents within the ocean. Where such specular reflection (or sun glint) occurs, the effect is generally so severe that no subsurface colour information can be retrieved. (However, sun-glint photographs of roughness patterns in the ocean surface have been qualitatively used to provide information on ocean circulation in much the same manner as imagery from synthetic and real aperture radars. Although these photographic data show great potential, they have not been practically adapted to image processing and, thus, are beyond the scope of this paper.)

The upwelled spectral radiance (or colour) of the ocean results from the backscatter of light energy from below the surface (less the spectral absorption). Measurements of upwelled radiance can provide information on the ocean's biological production and optical properties, and can act as tracers or descriptors of physical ocean events.

There is a practical depth limit from which the radiance is reflected. Gordon & McCluny (1975) call the depth at which the spectral radiance downwelled from the surface falls to 1/e of its initial energy 'the first optical attenuation length'. Approximately 90 % of the upwelled radiance is backscattered before this depth, and radiance from greater depths is negligible. For practical purposes, then, the upwelling spectral radiance of the ocean measured at the water surface may be considered to be the integration of the radiant backscatter and the spectral absorption within the first optical attenuation length.

For clear ocean waters, the attenuation length is normally 20 m or greater; in coastal areas, the attenuation length is typically 2–4 m. An inverse relation of the attenuation length – the diffuse attenuation coefficient, k – is normally used as a measurement of ocean radiance. Although k depends on the vertical distribution of scattering and absorption properties, it does not reveal the actual depth at which the optical variability occurs. In other words, it is possible to have the same water-leaving radiance at the surface for two separate areas whose optical properties are vertically very different. A relatively turbid, homogeneous water mass can display similar surface radiance to a clear water mass with a thin, turbid layer lying within the attenuation depth. Thus, any vertical variation in the optical properties that occurs in the first attenuation length can significantly influence the upwelling water-leaving radiation sensed by an ocean colour scanner.

The concentrations of phytoplankton pigments are largely responsible for the ocean's optical properties (these pigments are largely representative of chlorophyll-a and also include the phaeopigment concentration (Gordon *et al.* 1983). The spectral absorption ability of the pigment significantly affects upwelling radiance (Arnone & La Violette 1986). The strong correlation between this biological pigment and the water optical properties provides an excellent measurement of a region's productivity.

Atmospheric effects are also significant when sensing the ocean from space at visible wavelengths. As much as 90% of the radiance detected at satellite altitudes can be expected to be due to atmospheric scattering (Gordon 1978). Atmospheric scattering arises from scattering from gas molecules (called Rayleigh scattering) and from aerosols or particulates (called Mie scattering). The molecular composition of the atmosphere is relatively constant so that the removal of the Rayleigh component of the earth's radiance is not too complex. However, the aerosol effects, depending on particle concentration, composition and size distribution, are highly variable both spatially and temporally.

Limitations of infrared and visible satellite imagery

Most of the infrared and visible satellite imagery that can be used to define oceanographic events has limitations imposed by the blockage of clouds or by the integration of the ocean signal with the intervening atmospheric moisture and aerosols. Many methods have been formulated to minimize these effects; several will be discussed.

A second important limitation of infrared and visible satellite imagery is that these data represent the surface layers of the ocean. In the case of infrared imagery, the restriction is more severe because these data represent the 'skin' radiation of the ocean. In ocean regions that undergo extremely calm conditions and high solar radiation, the infrared imagery display abnormally high temperature readings, which are representative of extremely shallow surface heating rather than the temperature of the first metre of the water column (cf. the southern area of figure 1*a*). When this type of heating occurs, it is best to use night imagery of the region; at night solar heating is not present to create stratified conditions, and the shallow stratified layers left from the day are destroyed by dynamic mixing due to the ocean's normal thermal radiation. Most of the visible range imagery represents near-surface radiation (i.e. the portions of a visible image that shows upwelled radiance, not Fresnel reflection), so these data can be used to display frontal boundaries instead of the thermal imagery when shallow surface heating is a problem (cf. figure 2) and night-time infrared imagery is not applicable.

The satellites

Four satellite series now provide infrared and visible oceanographic data: *GOES* and *Meteosat* (in geosynchronous orbits), and NOAA and *Nimbus* (in polar orbit). Because the resolution of the *GOES* and *Meteosat* data is poor when compared to that of the NOAA and *Nimbus* data, this discussion will be limited to data from the latter.

The NOAA satellite series are operational satellites that normally carry the same sensor suite on each new spacecraft. Thus, image-analysis techniques useful for a specific sensor aboard the NOAA series can be used on data from almost any satellite in the series.

The NOAA advanced very high resolution radiometer (AVHRR) is a good example of an oceanographically useful sensor that has been aboard all NOAA satellites since *TIROS-N* (i.e. all NOAA satellites after *NOAA*-4). As the NOAA satellites are placed in Sun-synchronous orbits (figure 4), the AVHRR aboard each satellite can repeatedly detect the thermal (AVHRR IR)

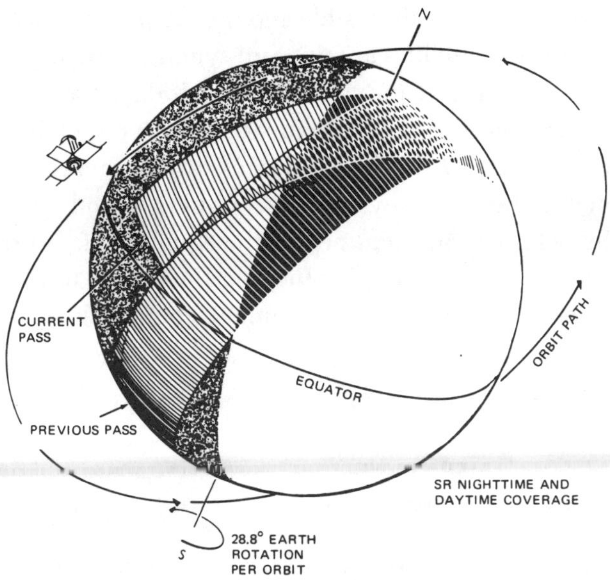

FIGURE 4. The Sun-synchronous, polar orbit of the NOAA satellite. Note the increase in the overlap of successive orbits near the poles.

and visible (AVHRR V) emissions of any region of the Earth at least twice each day. Thus, for cloud-free regions of the globe, 10-bit data from the AVHRR IR can convey descriptions of the surface thermal structure of the world's oceans approximately every 12 hours. A resolution of about 1 km and an absolute accuracy of ± 0.3 °C are obtained. Detailed information on the NOAA satellites and the AVHRR can be found in Schwalb (1978) and Hussey (1979).

Figure 1 is an example of the thermal gradient display made possible by using *NOAA*-7 AVHRR IR data. In this thermal image the equivalent notation of grey shades to temperature variation is the standard method used in satellite oceanography; i.e. higher temperatures are displayed as increasingly dark tones and lower temperatures as lighter tones, and the whitest feature is usually the cold tops of clouds. In most of the imagery presented here, a step function is used to display clouds as black. This highly useful display function de-emphasizes the seeming omnipresence of white clouds in an image, while allowing the presentation of sometimes subtle ocean thermal features.

For long-term operational purposes, the *Nimbus* research satellites are the least usable of the polar orbiters because the sensors on these satellites are usually one-time affairs; each new *Nimbus* spacecraft has a different suite of experimental sensors. The *Nimbus*-7 coastal zone colour scanner (CZCS), which can directly detect the variation and distribution of the ocean's chemical and biological material, is a good example. As figure 2 shows, however, the CZCS can indirectly provide information on currents and frontal features in the water.

Although the CZCS has proven to be a highly successful oceanographic sensor, it is not scheduled to be placed on any of the upcoming *Nimbus* spacecraft. Much of what will be said here, however, is applicable to other visible range sensors and provides good examples to follow when using similar sensors in the future.

The 8-bit data from the *Nimbus*-7 CZCS can detect subtle ocean colour changes by using five visible spectral bands selected for their ocean-viewing qualities (unlike *Landsat*, whose visible spectral bands are highly useful for land viewing but poor for looking at the oceans). These five channels are centred at 0.443, 0.520, 0.670, 0.750, and 1.150 μm, and have a resolution at nadir of 800 m. The first four channels are in the visible and the fifth is in the near infrared (reflective).

The polar orbits of the *Nimbus* satellites are not Sun-synchronous as are the NOAA satellites. Rather than passing over the earth at the same time each day, the orbit is such that a given region at 40° north or south is viewed on approximately three out of four days at around local noon. A description of the *Nimbus*-7 satellite and CZCS can be found in Hovis *et al.* (1980).

It is important to note that because they are polar orbiters, both NOAA and *Nimbus* provide far more coverage in the polar regions than nearer the equator (cf. figure 4 and table 1). This coverage can give a tremendous advantage in studying the oceans in these remote regions and allows such extremely useful techniques as multi-image compositing.

TABLE 1. *TIROS-N* ORBIT PREDICTIONS FOR OSLO, 30 OCTOBER 1980

pass	time of maximum elevation (Z)	time interval
1	02h27	—
2	04h08	01h41
3	05h29	01h41
4	10h45	06h16
5	12h24	01h39
6	15h46	03h22

SATELLITE IMAGE ANALYSIS TECHNIQUES

Basic computer techniques

Four basic image analysis techniques should be used to derive oceanographic information from satellite imagery: selective enhancement, geographic registration, absolute ocean radiation and multiple-image composition. Any number of subsets and combinations of these techniques are possible, such as edge enhancement or the ability to make loop movies of selected series of registered images.

Selective enhancements

Selective enhancement is one of the simplest and most useful tools of satellite image analysis techniques. Although the grey-scale wedge in the imagery is limited to a finite number of grey bands, these bands can be assigned to display selected ocean phenomena in the manner shown in figures 1*a*, *b* and 5*a*. Another method to emphasize the image's oceanic information utilizes edge enhancement, as shown in figure 1*c*. Colour wedges, as used in figure 2, can be used instead of the grey wedge. When the image is registered to a standard projection and the grey (or colour) shading is equivalent to absolute temperature or bio-optical values, the enhancement technique can be quantitatively, as well as qualitatively, useful.

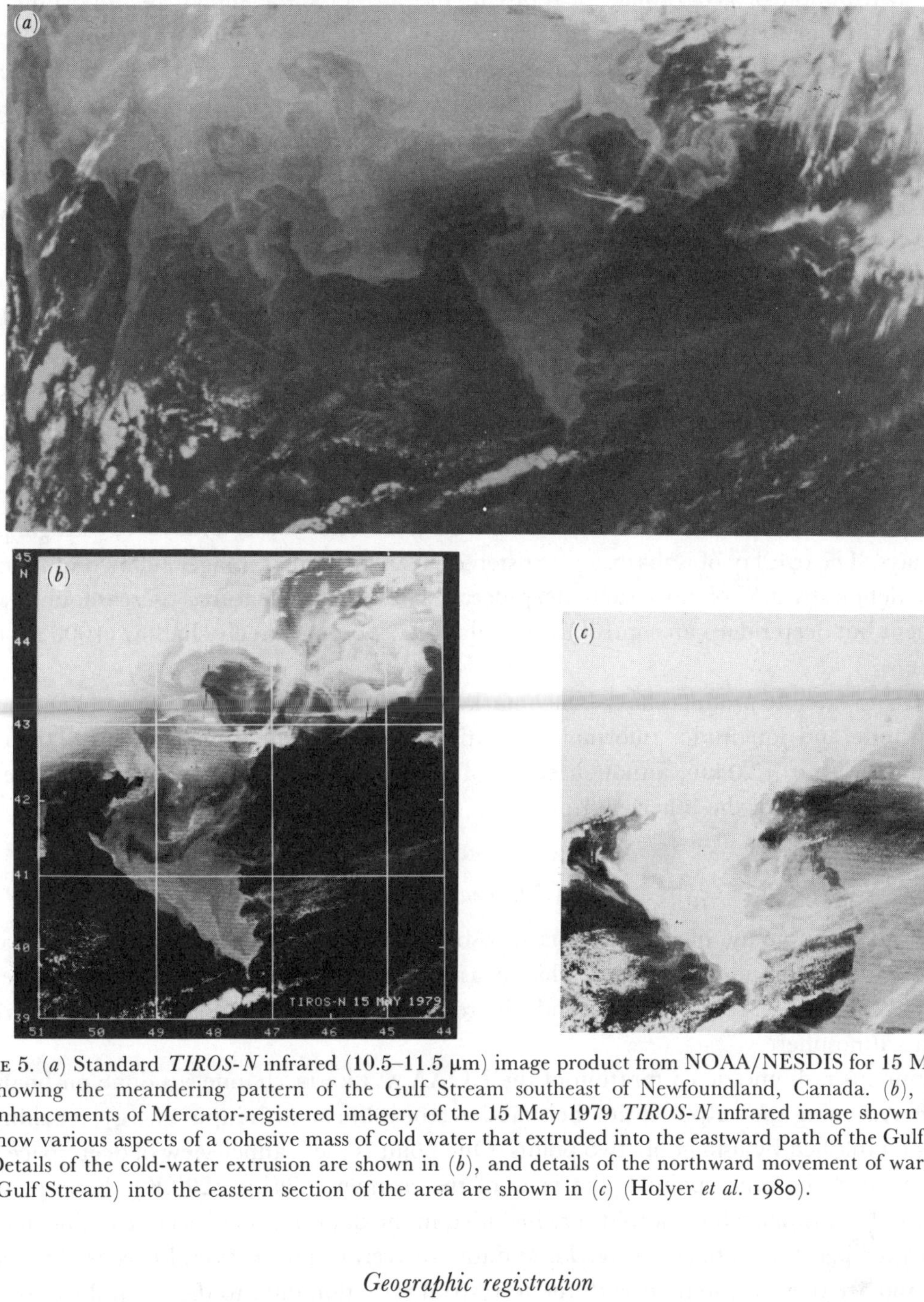

FIGURE 5. (*a*) Standard *TIROS-N* infrared (10.5–11.5 μm) image product from NOAA/NESDIS for 15 May 1979 showing the meandering pattern of the Gulf Stream southeast of Newfoundland, Canada. (*b*), (*c*) Two enhancements of Mercator-registered imagery of the 15 May 1979 *TIROS-N* infrared image shown in (*a*) to show various aspects of a cohesive mass of cold water that extruded into the eastward path of the Gulf Stream. Details of the cold-water extrusion are shown in (*b*), and details of the northward movement of warm water (Gulf Stream) into the eastern section of the area are shown in (*c*) (Holyer *et al.* 1980).

Geographic registration

Finding the accurate geographic location of ocean features seen in a satellite image is an important input to oceanographic analysis of the image, especially if the feature is to be correlated with data from other satellites, ships, or aircraft. Correlation is possible by using NOAA satellite data distributed in computer compatible tapes (CCT) by satellite data distribution centres, such as NOAA/NESDIS (U.S.) and CMS (France). These CCTs have geographic positions incorporated in a systematic fashion throughout the AVHRR data. Imagery constructed from these data has been shown to contain location accuracies with a standard deviation of 1.7 km about mean errors of 3.7 km (Clark & La Violette 1981).

The distribution of geographic locations in the AVHRR data allows an important image analysis technique to be used beyond geographically positioning a point on an image. It allows an image to be displayed in a standard map projection, such as Mercator or polar stereographic. Now, sets of NOAA imagery can be made that have a common projection and measurements made on temporal and spatial scales that were not easily done before. The method often used to warp NOAA-type data to various map projections uses two-dimensional, third-order polynomials derived from the latitude–longitude data on the CCT. This accuracy may be further improved if the ocean region has well-distributed landmarks. Almost all the imagery presented in this paper has been registered to a standard map projection. The exception is figure 5*a*, which shows how the unregistered image looks as received from the satellite, i.e. as a skewed cylindrical projection.

Figures 6 and 7 show thermal imagery of the Grand Banks and the Alboran Sea warped to Mercator projections. The warping was done with sufficient accuracy (± 1.0 km) to allow the overlay of data from other satellites, aircraft, and ships. For figure 7, the successive imagery was used to produce a time-lapse sequence, or cine loop, of ocean features with no noticeable misregistration jitter. The frames shown in figure 7 were a portion of a movie covering a period of ten days. The overlay of isobaths on registered NOAA thermal imagery shown in figure 8 is a good demonstration of how such deep-ocean bathymetric features as seamounts and a prominent but deep ridge can control the circulation of comparatively shallow (1000 m) ocean features.

The *GOES* satellite products have benchmark records that associate sample and line numbers with latitude and longitude. Informal evaluation has found these products to be generally accurate to within ± 20 km, although errors of several hundred kilometres are possible. The positional accuracy of the other environmental satellites has not been evaluated.

Absolute ocean radiation

If the quantitative use of satellite data is to be expanded, an accurate measurement of the absolute ocean radiation is needed. This measurement requires, first, the radiometric calibration of the satellite sensor and, second, the removal of the portion of the signal that results from the atmosphere.

Continuous radiometric calibration of the NOAA AVHRR IR channels is done on-board the satellite. Assuming the output of the thermal detector is linear with input energy, a calibration curve is established by using only two points. One point is the scanner view of deep space (zero radiance), and the other point is the view of the scanner housing (290 K). Four platinum resistance thermistors, whose outputs are included in the data stream of each scan line, monitor the exact temperature of the housing. These data are averaged over several sweeps of the sensor mirror and are then transmitted with the terrestrial radiation data to the ground station. The resultant curve provides a dynamic calibration that is used to track and compensate for any drift in the performance of the detector and associated electronics. CCTs received from central satellite data collection stations, such as NOAA/NESDIS, normally contain these calibration data.

The oceanic signal lost due to the overriding signal of the atmosphere places a strong limitation on the oceanographic use of infrared and visible imagery. With infrared imagery, this problem is especially prevalent in data taken at mid- and low latitudes in summer.

FIGURE 6. Mercator-registered *TIROS-N* infrared (10.5–11.5 μm) imagery of the waters of the Grand Banks southeast of Newfoundland, Canada, for four successive days in May 1979 (the series includes the image in figure 5). The imagery shows the movement of frontal shear waves moving along the boundaries of a cohesive mass of cold water that had been extruded into the eastward path of the Gulf Stream. The shear waves were active on both the western and eastern fronts of the cold-water mass and moved at phase speeds of approximately 70 cm s^{-1}, a period of approximately 0.83 days and a wavelength of 50 km. Movement during a period of 12 hours is marked by arrows. A comparison of these derived phase speeds with *in situ* measured current speeds shows the waves were moving at nearly the same speed as the cold-water currents. The imagery shows that rapid movement within the ridge feature was also measurable. Tracking the displacement of several cold filaments within the cold water over several days shows they averaged 70 cm s^{-1}. These displacements are similar to velocities measured by satellite drifter buoys tracked during the same period. The imagery has not been atmospherically corrected (La Violette 1983).

Of several possible solutions, the most popular is the multichannel single-satellite approach (McClain *et al.* 1977).

The NOAA AVHRR IR has three channels centred at the 3.7, 11 and 12 μm atmospheric windows. The spectral bands thus have relatively high transmittances with regard to middle- and far-infrared spectral range energy emitted by the ocean surface. As mentioned earlier, the most significant atmospheric absorption constituents in these regions are water vapour and aerosols. However, the amount of absorption varies for each spectral region. The 3.7 μm channel has a transmittance of 90 %, the 11 μm channel has a transmittance of 75 %, and the

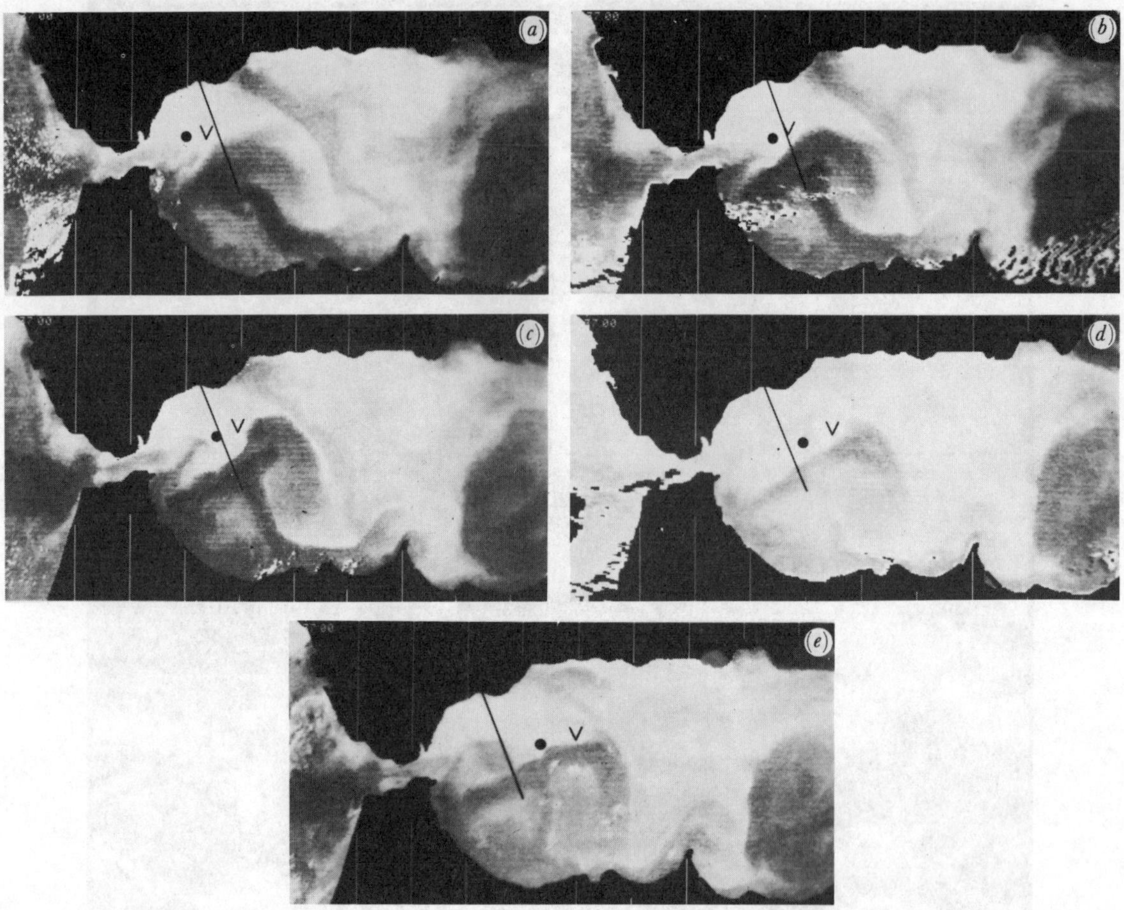

FIGURE 7. *NOAA*-7 AVHRR-IR imagery of the Alboran Sea collected as part of the Donde Va? experiment during October 1982. The dot and the 'V' in the images show the advection of two submesoscale cold features about the permanent gyre located in the Alboran Sea. The black line represents a line of current meter moorings positioned during the Donde Va? experiment. A close examination of the imagery shows that other unmarked features were also being advected and that their generation was probably tide related. Continuous monitoring of these cold-water features was possible because of the twice-daily (about every 12 h) spacing of the *NOAA*-7 overpasses. After registration to a Mercator projection (accuracy ± 1 km) and an atmospheric correction to arrive at absolute temperatures, analysis of the displacement of the cold features in successive images shows their apparent origin east of Gibraltar, their average speeds of 40 cm s^{-1} around the gyre, and their apparent entrailment into the incoming Atlantic water east of Gibraltar (La Violette 1984). (*a*) 14h32 GMT 6 October; (*b*) 02h56 GMT 7 October; (*c*) 14h22 GMT 7 October; (*d*) 02h44 GMT 8 October; (*e*) 14h08 GMT 8 October.

12 μm channel transmittance is 80%. (Although limited to night use, one of the preferred channels to use because of its high transmittance rate is the 3.7 μm channel; however, high noise levels in this channel in all of the post-*TIROS-N* AVHRR IRS have precluded any real use of a three-channel algorithm. Until now only a two-channel algorithm has been used.)

These variations in infrared transmittance for the same ocean scene are the basis of the multiple-channel algorithm. The algorithm uses a correlation made by subtracting the effects of one channel from another; the remainder is a function of the amount of moisture present in the atmosphere. The two-channel equation takes the form

$$\text{MCSST} = A_1(T_{11}) + A_2(T_{12}) + C,$$

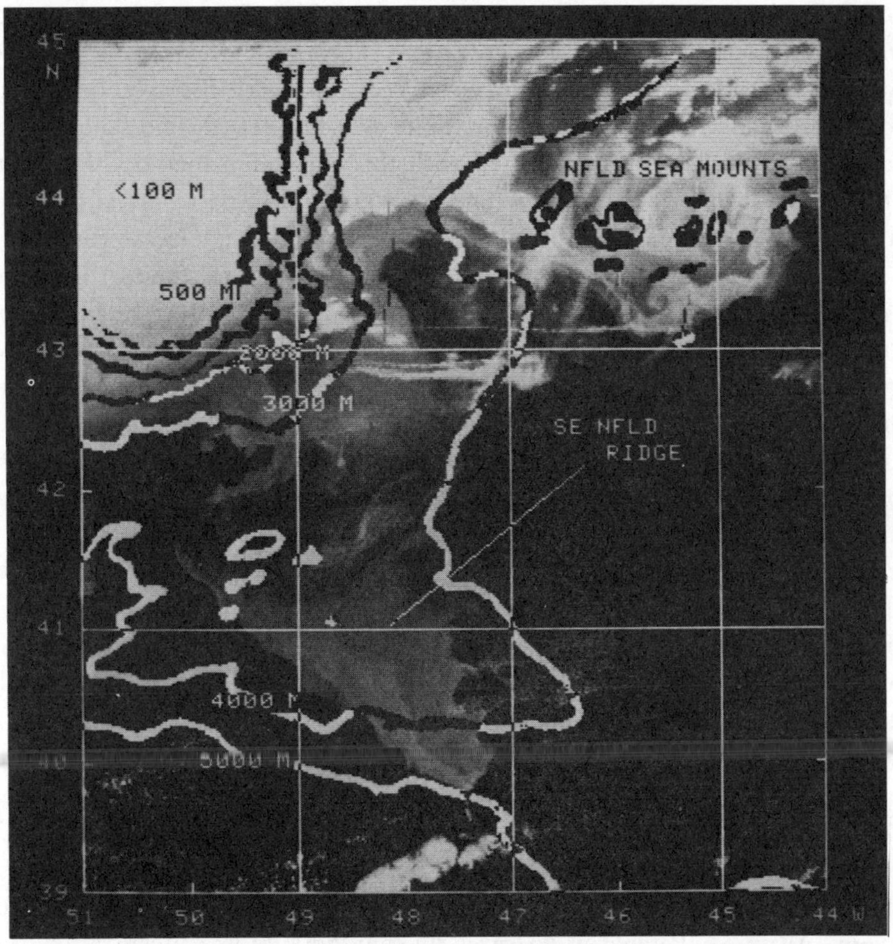

FIGURE 8. *TIROS-N* image for 15 May 1979 overlain with the 500, 1000, 2000, 3000, 4000, and 5000 m isobaths (Holyer *et al.* 1980).

where MCSST is the multichannel sea surface temperature, T_{11} is the measured radiative temperature in the 11 μm channel, T_{12} is the measured radiative temperature in the 12 μm channel, A_1, A_2 and C are coefficients derived from a best fit to empirical surface data.

If the greatest atmospheric problem with infrared data of the ocean is water vapour, with aerosols playing a minor role, then the reverse is true of visible range data. Aerosols are not only the major absorber–scatterer of the visible atmospheric signal, but their effect often results in a loss of almost 90% of the ocean signal.

The most widely accepted method for correcting the effects of aerosols in CZCS data requires extensive computer manipulation involving assumptions about the scattering properties of the local atmosphere and ocean. The method uses a mathematical atmospheric model based on single-scattering theory (Gordon & Clark 1980*a*)

$$L_{W(\lambda)} = L(\lambda) - L_R(\lambda) - (F(\lambda)/F(670))\,(L(670) - L_R(670)), \tag{7}$$

where $L_{W(\lambda)}$ is the upwelling radiance from the ocean at wavelength λ, $L(\lambda)$ is the radiance observed from space, $L_R(\lambda)$ is the calculated Rayleigh scattering radiance of the atmosphere, $F(\lambda)$ is solar flux.

The Gordon & Clark (1980*a*, *b*) method assumes the CZCS 0.670 μm channel can be used as a measure of the atmospheric aerosol concentration. Through a weighted subtraction of this channel with appropriate subtraction of Rayleigh scattering, the effective upwelling radiance from the ocean surface can be computed for the CZCS 0.443, 0.520, and 0.550 μm channels.

However, several problems are associated with this correction method. The first problem involves the proper selection of the Ångström exponent. The Ångström exponent relates the wavelength-dependence of the aerosol optical thickness and single scattering albedo of the aerosols. (For a more complete explanation, see Gordon & Clark 1980*a*.) Gordon & Clark (1981) suggest that the radiance values of clear water in any CZCS image be used to obtain this value. This technique, however, requires knowing the location of clear-water areas within the CZCS image. Furthermore, the assumption is made that the Ångström coefficient over the clear water is the same throughout the image. If this assumption is not valid, the selected Ångström exponent value for one part of the image will produce erroneous values when applied to other portions of the image (Arnone 1983*a*, *b*).

To overcome this difficulty, Arnone & La Violette (1984) developed an interactive analysis procedure that defines the optimum coefficient for the entire image scene. This method was used for the CZCS images presented in figures 2 and 11.

A second problem with the Gordon & Clark (1980*a*) correction procedure arises from the assumption that water has zero upwelling radiance at 0.670 μm. This assumption is logical for open ocean and shelf-type water masses where minimal concentration of suspended sediments occurs. However, significant upwelled radiance at 0.670 μm occurs in turbid coastal waters (Smith & Wilson 1981; Arnone 1983*a*, *b*). For these coastal areas, an underestimation of the absolute upwelling radiance in the correct CZCS channels will result if this method is used.

Arnone & La Violette (1986) used the ratio of the absolute upwelling radiance for the 0.443 and 0.550 μm channels (i.e. 0.443/0.5500 to derive both the diffuse attenuation coefficient k at 0.490 μm (Austin & Petzold 1980) and the phytoplankton concentration (Gordon & Clark 1980*a*)). This method implies an inherent relation between k and phytoplankton concentration.

Multiple-image composition

The problem of clouds blocking the infrared satellite sensor's view of the ocean has long plagued oceanographers. Fortunately, ocean events change more slowly than atmospheric events. By using this fact as a basis, an analysis technique that minimizes cloud and atmospheric contamination can be arranged. This method composites the highest temperature values for a given region from infrared data collected over a period of several days (La Violette & Chabot 1969). The technique applies to any accurately registered thermal IR imagery using a common projection (figure 9).

One limitation of this method is that short-term effects are smeared in the final composite. Also, the composite is biased toward the period when the area was cloud-free. Like any such data restriction problems, these limitations can be taken into account and weighted according to the desired objective. Essentially, the user is faced with the problem of using the smallest number of satellite passes that will give the most cloud-free conditions and do the least smearing of ocean features.

The technique can be used most advantageously with ocean data from the higher latitudes. Here, the increased number of overflights each day by the polar-orbiting satellites allow composites to be made of several passes in a given day (table 1). As many of the storm systems

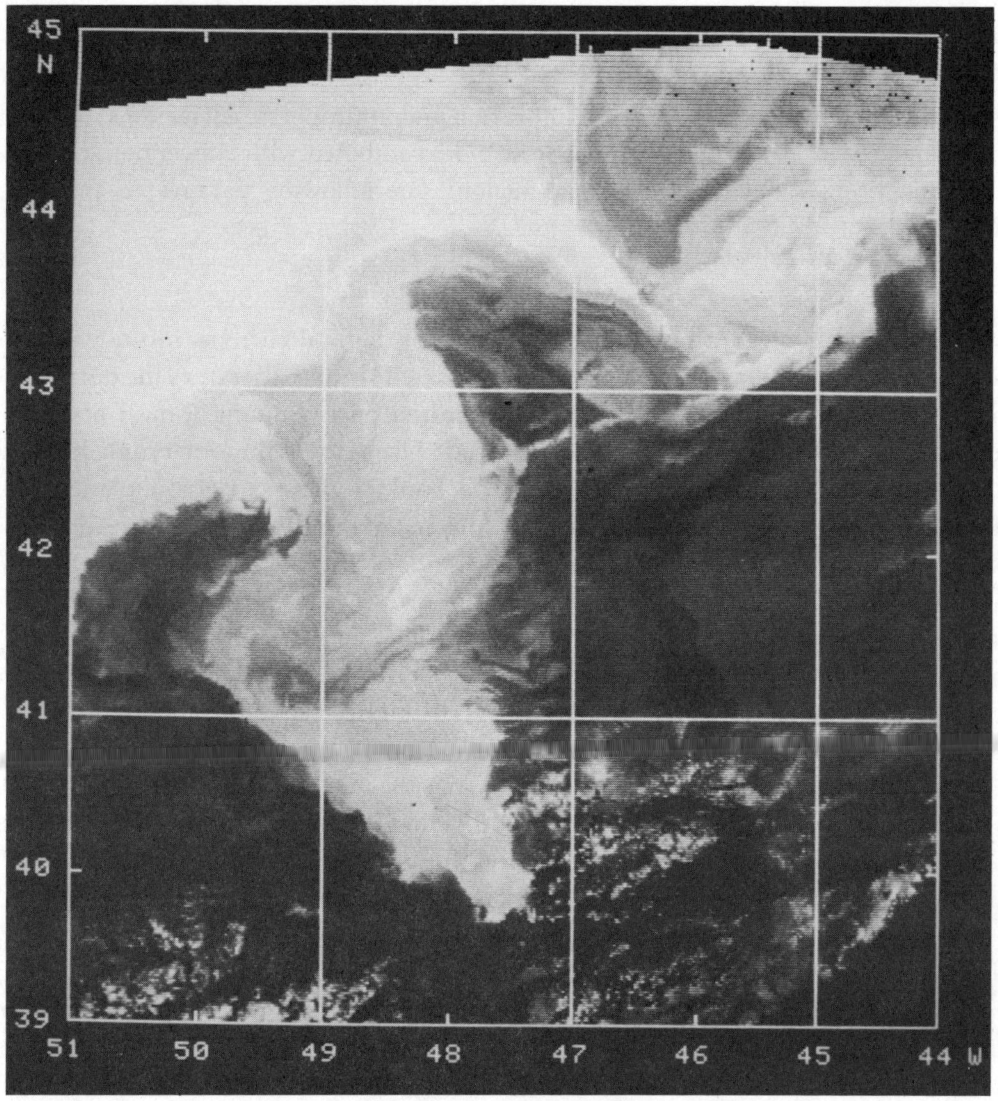

FIGURE 9. The four *TIROS-N* images in figure 6 merged into a composite designed to minimize the effect of clouds. The composite was formed by comparing the temperatures for four days at each pixel location and retaining the warmest temperature at each address for those four days (Holyer *et al.* 1980).

in the polar regions are fast moving, the chances are increased of a satellite passing over a cloud-free period during a given day. Compositing the several passes that occur in 1 or 2 days can minimize the amount of ocean change that may have taken place.

An expansion of the compositing method developed by La Violette & Chabot (1969) uses a sliding composite technique (Holyer *et al.* 1980). In this method, a selected number of days are composited to form the first of a series of images. The next composite image uses the same number of days, but drops the first day and adds the next chronological day to the series. The resulting composite image is, thus, one day removed from the first composite image. The method is continued until a satisfactory series that can be made into a loop cine is made. One potential application of this technique is to make loop movies of satellite imagery to compare with cine loops derived from numerical models.

Examples of oceanographic analysis using processed satellite image and conventional data

If the basic functions (i.e. registration, calibration and atmospheric corrections) are properly done, interactively processed satellite imagery can be combined with conventional ocean data in the oceanographic analysis of an ocean region. The following paragraphs present some typical examples.

The Grand Banks experiment

This experiment involved a two-year ocean study of thermal currents in the waters of the Grand Banks of Newfoundland. The field work used satellites and aircraft as the data collection platforms. The May 1979 portion of the study centred on examining a mass of cold water 350 km long and 100 km wide that had been extruded from the Labrador Front. Much of the following was taken from the results of the Grand Banks experiment (La Violette 1981; La Violette 1983). The satellite used was *TIROS-N*, the first of the NOAA satellites equipped with AVHRR.

On 16 May, a frontal survey, which consisted of a research aircraft flying sorties along a right-angle cut across the cold feature, was made. The aircraft operation flight lines were determined by 15 May *TIROS-N* imagery that had been registered and enhanced before the flight. Analysis of the aircraft data produced a vertical cross section of the atmospheric and oceanic conditions in one portion of the study region on 16 May.

To relate this comparatively narrow view to the overall regional conditions, *TIROS-N* infrared data were registered and enhanced, and a plot of the aircraft survey track was overlaid on the image shown in figure 10*a*. The figure also shows an ocean cross section produced from aircraft expendable bathythermographs (AXBTS). The cross section indicates that the cold water in the satellite image is not the surface spreading of windblown, low-density water, but is a large major extrusion that extends into the water column deeper than 350 m. Surface currents were plotted by using the rate and direction of drifting sonobuoys dropped from the aircraft (figure 10*b*).

The regional circulation over a period of four days (including the 16 May aircraft survey) was also studied using *TIROS-N* imagery (figure 6). Because of the registration precision, day-to-day movements along the front could be studied and accurately measured. Within the cold water to the north, filaments of cold water are shown to be rapidly moving about the feature. Although the day-to-day shape of the filaments became distorted, they showed a surprising ability to retain their identity.

The Donde Va? experiment

The purpose of Donde Va? was to define the variability and the causes of the surface circulation in the Alboran Sea (La Violette *et al.* 1982; Kinder 1983; The Donde Va? Group 1984). In addition to the satellite data, the experiment involved aircraft, ships and moored buoy data.

The twice-daily passes of the *NOAA*-7 over the study area permitted almost continuous monitoring of the surface thermal changes that occurred in the Alboran Sea during the period when intensive air and sea measurements were being conducted (October 1982). The satellite infrared imagery showed that a number of submesoscale cold-water features were being

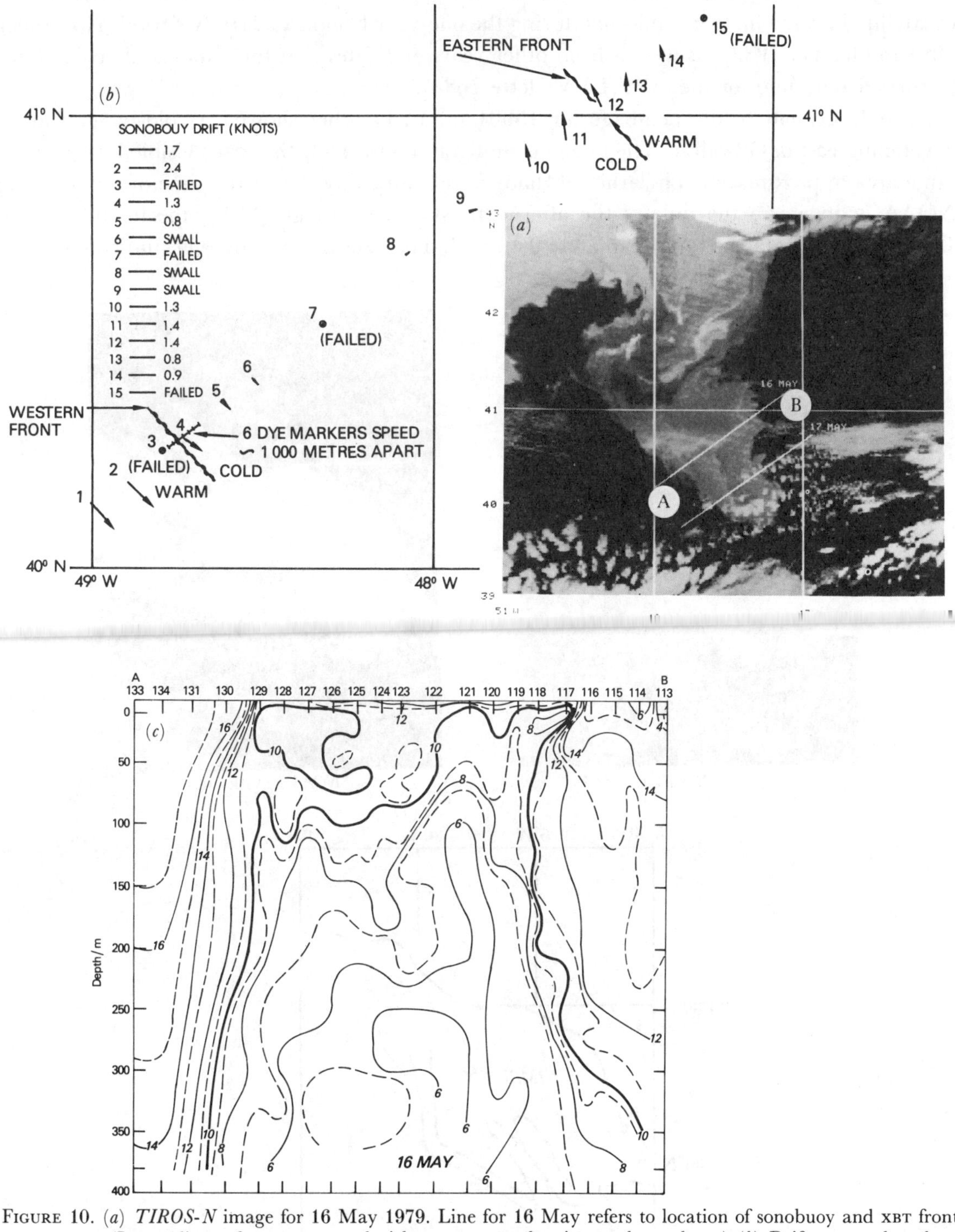

FIGURE 10. (*a*) *TIROS-N* image for 16 May 1979. Line for 16 May refers to location of sonobuoy and XBT front survey. (Lower line refers to a second airborne survey that is not shown here.) (*b*) Drift vectors based on sonobuoy data. (*c*) Temperature vertical cross section based on airborne XBT data. Numbers between A and B refer to station numbers and positions of airborne XBTS (La Violette 1983).

advected about a large gyre that dominated the western Alboran (figure 7). The examination of satellite infrared imagery collected during the one-year Donde Va? study period (November 1981 to October 1982), as well as from other years, indicates that these mesoscale cold-water features are normal for the gyre (La Violette 1984).

An enlargement of the IR image for 15h01 on 12 October showed a cold-water feature developing east of Gibraltar. The image is shown in figure 11 with a CZCS visible image of the same area approximately 3 h earlier (11h55) of the same day. From the time of the morning NOAA IR image to the time of the afternoon NOAA IR image (11 h), the feature moved 16.7 km at an average of approximately 0.4 m s^{-1}. The CZCS image, however, indicated that a

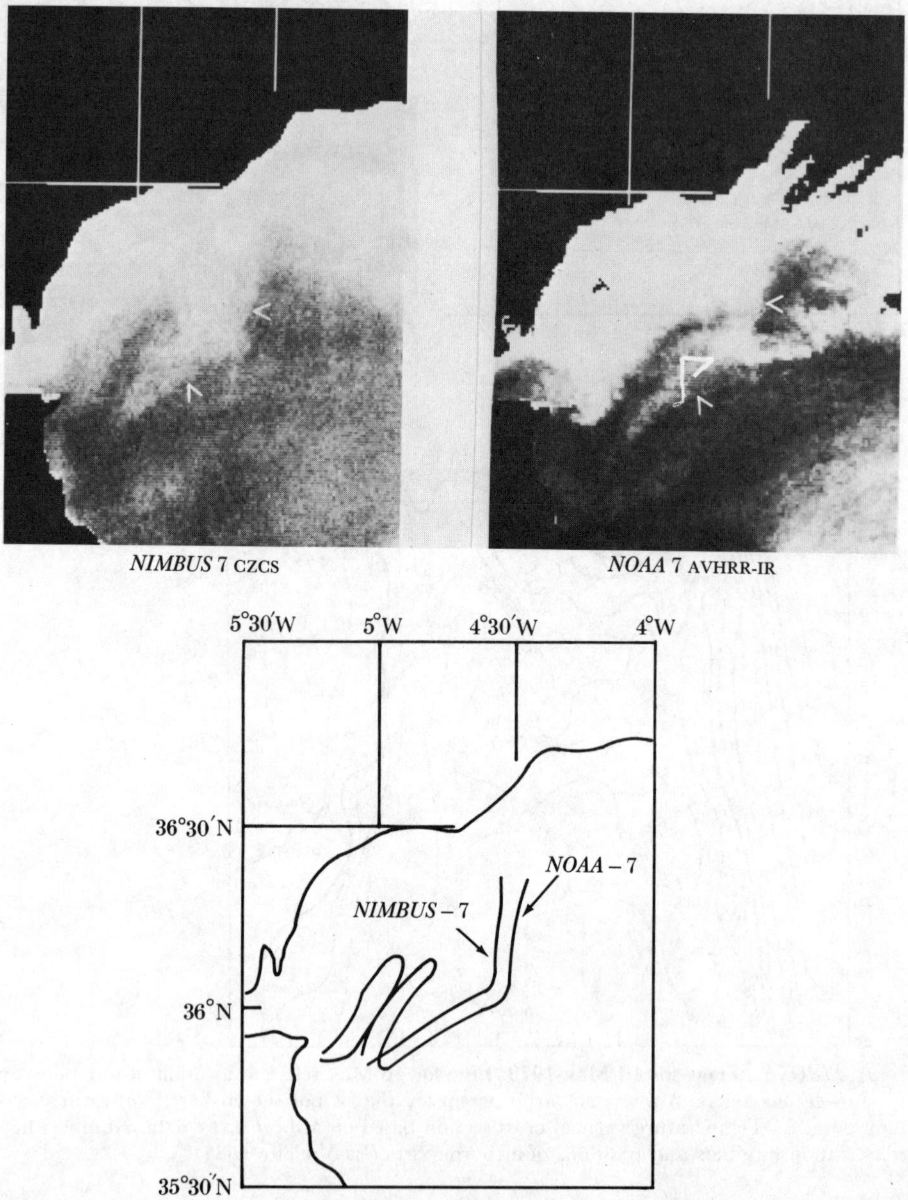

FIGURE 11. The detection of tide-current-related, short-term movement of a cold-water feature in the Alboran Sea by using Mercator-registered *Nimbus* CZCS and *NOAA*-7 AVHRR-IR imagery. The movement of approximately 9 km took place between 11h55 (*Nimbus* image) and 15h01 (NOAA image) on 12 October 1982 (La Violette 1984).

movement of approximately 9 km occurred in the time between the *Nimbus* noon pass and NOAA afternoon pass. This equates to a movement of 0.8 m s^{-1}, or twice the rate of movement for the period of 11 h between the two NOAA images.

When the data were compared with the tide tables of the region, the time of the spurt in speed of the feature was found to correspond to the time of the maximum eastward component of the tidal current. This and other satellite and conventional data indicated a strong influence of the tidal currents in the circulation of the upper 200 m of water in the western Alboran Sea (La Violette & Lacombe 1987).

In a study of the drift rate of sonobuoys dropped from a research aircraft during 9 through

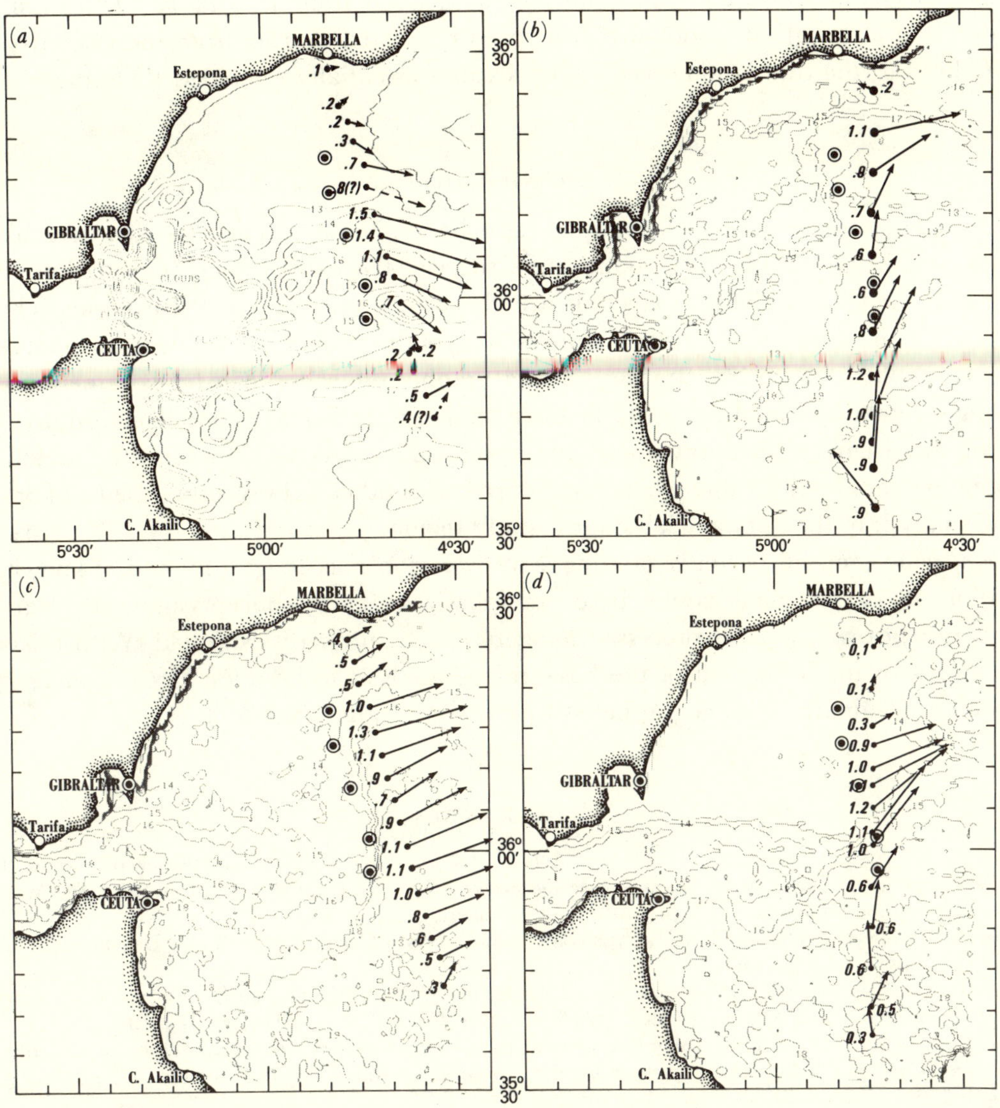

FIGURE 12. The positions of short-term sonobuoy drift movements (in metres per second) in relation to the sea-surface temperature field devised from *NOAA*-7 AVHRR IR data during the Donde Va? experiment. The circled dots refer to a line of current meter moorings also in position at that time. This is an example of how field-collected data can be integrated with accurately Mercator-registered and atmospherically corrected satellite imagery (after La Violette 1986). (*a*) Start 10h25, finish 15h15, average time 04h30 min; (*b*) start 09h17, finish 13h21, average time 03h48 min; (*c*) start 09h58, finish 15h58, average time 5h38 min; (*d*) start 09h30, finish 13h46, average time 3h45 min.

18 October, La Violette (1986) showed that surface currents associated with the incoming Atlantic water flowed at speeds equal to or greater than 1.0 m s^{-1} and that, on the average, the greatest speeds (i.e. speeds of 1.2 m s^{-1}) were limited to a band less than 16 km wide centred approximately between the 17 and 18 °C surface isotherm. Of interest is the fact that the band of fast-moving surface water flowed just inside the periphery of the advected cold-water features shown in the satellite imagery (figure 12).

Comparing the average drift speed of the sonobuoys (*ca.* 1.2 m s^{-1}) at the periphery with the average advection speed of the cold-water feature (0.4 m s^{-1}) gave an indication of the differences between the surface currents and the translation speed of the feature around the gyre. Comparison of current speeds derived from sonobuoys dropped in the southeastern part of the gyre on 18 October and the translation speeds derived from 18 and 19 October imagery revealed that although the sonobuoy drift speed decreased sharply from the velocities found near Gibraltar, the translation speed of the features essentially remained the same.

Conclusions

Certain basic image analysis techniques must be applied to satellite data so that this form of data will be a reliable and quantitative source of oceanographic information. Proper registration and atmospheric correction are the most important of these basic techniques (especially with visible range data where up to 90 % of the signal may be due to the intervention of the atmosphere). Also, registration to a set graphic projection can be extremely useful in allowing several days of data to be studied as a temporal, as well as a spatial, continuity. When the data are accurately calibrated and atmospherically corrected, however, the usefulness of the data expands considerably when coregistered with other data sets and used as part of an overall analysis of the data available for a study region.

The specific concepts of understanding a study region's oceanography and of knowing the limitations of satellite data cannot be overemphasized. These concepts must be clear to an investigator before computer-processed imagery can be properly exploited. With this knowledge and with the utilization of the basic image analysis techniques, satellite imagery can provide an imaginative oceanographer with a powerful analytical tool.

References

Arnone, R. A. 1983*a* Evaluation of CZCS and *LANDSAT* for coastal optical and water properties. In *17th International Symposium on Remote Sensing of the Environment, San Jose, Costa Rica, April 23–30 1980*. Ann Arbor, Michigan: Environmental Research Institute of Michigan.

Arnone, R. A. 1983*b* Water optics of the Mississippi sound. Naval Ocean Research and Development Activity, NSTL, Mississippi, report 63.

Arnone, R. A. & La Violette, P. E. 1984 A method of selecting optimal Ångstrom coefficients to obtain quantitative ocean color data from Nimbus 7 CZCS. *SPIE*, vol. 489 (Ocean Optics VII), pp. 187–194.

Arnone, R. A. & La Violette, P. E. 1986 Bio- and chem-optical variability of the shear eddies associated with the African Current off Algeria as assessed by the *Nimbus*-7 coastal zone color scanner. *J. geophys. Res.* **91**, 2351–2364.

Austin, R. W. & Petzold, T. J. 1980 The determination of the diffuse attenuation coefficient of sea water using the coastal zone colour scanner. In *Oceanography from space* (ed. J. F. R. Gower), p. 239. New York: Plenum Press.

Clark, R. & La Violette, P. E. 1981 Detecting the movement of oceanic fronts using registered TIROS-N imagery. *Geophys. Res. Lett.* **8**, 229–232.

Donde Va? Group 1984 Donde Va? An oceanographic experiment in the Alboran Sea. *Eos, Wash.* **65** (36), 682–683.

Gordon, H. R. 1978 Removal of atmospheric effects from satellite imagery of the oceans. *Appl. Opt.* **17**, 1631–1636.
Gordon, H. R. & Clark, D. K. 1980*a* Atmospheric effects in remote sensing of phytoplankton pigments. *Bound.-Layer Met.* **18**, 299–313.
Gordon, H. R. & Clark, D. K. 1980*b* Remote sensing optical properties of a stratified ocean: an improved interpretation. *Appl. Opt.* **19**, 3428.
Gordon, H. R. & Clark, D. K. 1981 Clear water radiances for atmospheric correction of coastal zone color scanner imagery. *Appl. Opt.* **20**, 4175–4180.
Gordon, H. R. & McCluny, W. R. 1975 Estimation of the depth of sunlight penetration in the sea in remote sensing. *Appl. Opt.* **14**, 413–446.
Gordon, H. R., Brown, V. W., Brown, O. B., Evans, R. H. & Clark, D. K. 1983 *Nimbus*-7 CZCS: reduction of its radiometric sensibility with time. *Appl. Opt.* **22**, 3929.
Holter, M. 1970 Imaging with nonphotographic sensors. In *Remote sensing with special reference to agriculture and forestry*, ch. 3. Washington, D.C.: National Academy of Sciences.
Holyer, R. J., La Violette, P. E. & Clark, J. R. 1980 Satellite oceanography research, development, and technology transfer. In *Proc. 14th Int. Symposium on Remote Sensing of the Environment, San Jose, Costa Rica, April 23–30 1980*. Ann Arbor, Michigan: Environmental Research Institute of Michigan.
Hovis, W. A., Clark, D. K., Anderson, F., Austin, R. W., Wilson, W. A., Butler, E. I., Ball, D., Gordon, H. R., Mueller, J. L., El-Sayed, S. F., Sturm, B., Wrigley, R. C. & Yentsch, C. 1980 *Nimbus*-7 Coastal Zone Color Scanner: system description and initial imagery. *Science, Wash.* **210**, 60–63.
Hussey, W. J. 1979 The *TIROS-N/NOAA* operational satellite system. National Environmental Satellite Service/*NOAA*, Washington, D.C., *NOAA NESS* tech. memo 95, 35 pp.
Kinder, T. H. 1983 Donde Va? An oceanographic experiment near the Strait of Gibraltar. In *Proceedings of IAPSO/ONR/NEFKD Workshop on Straits*. Copenhagen: Inst. Phys. Ocean.
La Violette, P. E. 1981 Variations in the frontal structure of the southeastern grand banks. Naval Ocean Research and Development Activity, NSTL-Bay St. Louis, Mississippi, report 87, 48 pp.
La Violette, P. E. 1983 The Grand Banks experiment: a satellite/aircraft/ship experiment to explore the ability of specialized radars to define ocean fronts. Naval Ocean Research and Development Activity, NSTL-Bay St Louis, Mississippi, report 49, 126 pp.
La Violette, P. E. 1984 The advection of the submesoscale thermal features in the Alboran Sea gyre. *J. phys. Oceanogr.* **14**, 550–565.
La Violette, P. E. 1986 Short-term measurements of the velocity of currents associated with the Alboran Sea Gyre during Donde Va?. *J. phys. Oceanogr.* **16**, 262–279.
La Violette, P. E. & Chabot, P. L. 1969 A method of eliminating cloud interference in satellite studies of sea surface temperature. *Deep Sea Res.* **16**, 534–547.
La Violette, P. E., Kinder, T. H., Preller, R. & Hurlburt, H. E. 1982 Donde Va?: a mesoscale flow dynamics experiment in the Strait of Gibraltar and Alboran Sea. In *Proc. 28th Congress and Plenary Assembly of CEISM, Cannes*, 2–11 December.
La Violette, P. E. & Lacombe, H. 1988 Tidal-induced pulses in the flow through the Strait of Gibraltar. *Oceanol. Acta*. (In the press.)
McClain, E. P., Pichel, W. G., Walton, C. C., Ahmad, Z. & Sutton, J. 1977 Multichannel improvements to satellite-derived global sea surface temperatures. *Adv. Space Res.* **2**, 43–47.
Schwalb, A. 1978 Modified revisions of the improved TIROS Operations (ITOS D-G). National Environmental Satellite Service/*NOAA*, Washington, D.C., *NOAA NESS* tech. memo *NESS* 35, 48 pp.
Smith, R. C. & Wilson, W. H. 1981 Ship and satellite bio-optical research in the California bight. In *Oceanography from space* (ed. J. F. R. Gower), p. 281. New York: Plenum Press.

Discussion

J.-P. A. L. Muller (*Department of Photogrammetry and Surveying, University College London, U.K.*). How far has the understanding of satellite oceanographic images advanced the knowledge of what the flows seen in satellite image sequences actually represent in terms of surface currents and circulation? In particular, are there any quantitative models which reproduce the satellite flow fields observed?

P. E. La Violette. As I have tried to emphasize, the features seen in the imagery are tracers of the circulation and are not direct measurements *per se*. If used judiciously in this regard, then the direction of currents (but not their speed) and the displacement (including rate of displacement) and variation of circulation features can be accurately inferred. For example, the monitoring of satellite-derived turbulent jets and eddies associated with the California current system have identified these features as crucial to the cross-shore transport. Their variance seasonally has been shown to be a direct result of seasonal changes in the velocity structure of the California current (see, for example, Kosro & Huyer 1986). Similar work with satellite imagery has been done to derive the circulation structure of multiple dipole eddies in the Alaska coastal current (Ahlnas *et al.* 1987). The ability of the movement of circulation features in successive imagery to act as lagrangian drifters is a more direct application of these imagery studies (Kelly 1983, 1985). There are problems with this in that the events are moving vertically as well as horizontally and that translation speeds of the features can be confused with current speed. Much of the early work by this method (see, for example, La Violette 1984; Vastano & Borders 1984; Vastano & Reid 1985) was done manually and is operator subjective. Objective computer techniques that use sea surface velocities are just now starting to be developed (see, for example, Emery *et al.* 1986). This developing methodology, used with constraints such as I have tried to emphasize in my paper (this symposium) promises to become an extremely powerful tool in satellite ocean research.

References

Ahlnas, K., Royer, T. H. & George, T. H. 1987 *J. geophys. Res.* **92** (C12), 13041–13047.
Emery, W. J., Thomas, A. C., Collins, M. J., Crawford, W. R. & Mackas, D. L. 1986 *J. geophys. Res.* **91**, 12865–12879.
Kosro, P. J. & Huyer, A. 1986 In *CTD and velocity surveys of seaward jets of northern California, July 1981 and 1982* (*J. geophys. Res.* **91**), pp. 7680–7690.
La Violette, P. E. 1984 *J. phys. Oceanogr.* **14**, 450–505.
Vastano, A. C. & Borders, S. E. 1984 *Remote Sens. Environ.* **16**, 87–90.
Vastano, A. C. & Reid, R. O. 1985 *J. atmos. Technol.* **2**, 393–400.

Phil. Trans. R. Soc. Lond. A **324**, 347–363 (1988)
Printed in Great Britain

A system for the analysis and comparison of oceanographic data from satellite-borne and *in-situ* sensors

BY M. D. MILNES[1] AND J. KEELING[2]

[1] *Logica Space and Defence Systems Ltd, Cobham, Surrey KT11 3LX, U.K.*
[2] *Admiralty Research Establishment, Portland, Dorset DT5 2J5, U.K.*

[Plate 1]

ODESSA is an Oceanographic Database and Environmental Satellite System Application developed for ARE (Portland) by Logica Space and Defence Systems Limited, under Ministry of Defence contract.

It is possible to access online or historic satellite images, which may be processed and displayed in a variety of formats. The oceanographic database contains data as a function of depth that have been derived from a wide range of instruments, geographic areas and times. This database may be accessed and the data processed and displayed in a number of ways, some of which can be directly overlaid and compared with the processed satellite images.

The system was designed to be flexible, with an emphasis on the man–machine interface, making extensive use of menus and graphic input devices. The flexibility allows for easy expansion to include further processing and display facilities.

Examples of output displays are presented to show the application of the ODESSA system in oceanographic modelling studies and in relating sea-surface temperatures to the three-dimensional water-column structure.

PART I. SYSTEM OVERVIEW

1. INTRODUCTION

An important area in the field of remote sensing is the interpretation of images. It is possible to analyse satellite images individually, in isolation; however, more useful information can be obtained by making use of additional data. These data could be taken from earlier remote observations of the same location, or from *in situ* data, obtained by more conventional means.

The ODESSA system provides the facilities to enable such additional information to be used. In particular, the current *in situ* data set consists of oceanographic data. This requires a considerable amount of processing of satellite images and of oceanographic data before meaningful comparisons between the two data sets can be made.

It is this processing that is described below, from the origins of the algorithms, through the design, to the implementation of the system.

Following this description, and a review of the future potential uses, a specific example of the current application at the Admiralty Research Establishment will be presented.

2. Background

The system is based on algorithms obtained from the NATO Oceanographic Research Centre (SACLANTCEN) at La Spezia, Italy. These algorithms were developed over a number of years, from 1974 onwards. However, there have been many changes and additions since then, some of which were based on work done for ARE by Logica, investigating the feasibility of the use of infrared satellite images for detecting various oceanographic features, an area of considerable interest in recent years (Robinson 1985). Following this study, ARE placed a contract with Logica to provide a facility for the continuation of this research, and to extend it to include analysis of historical *in situ* oceanographic data.

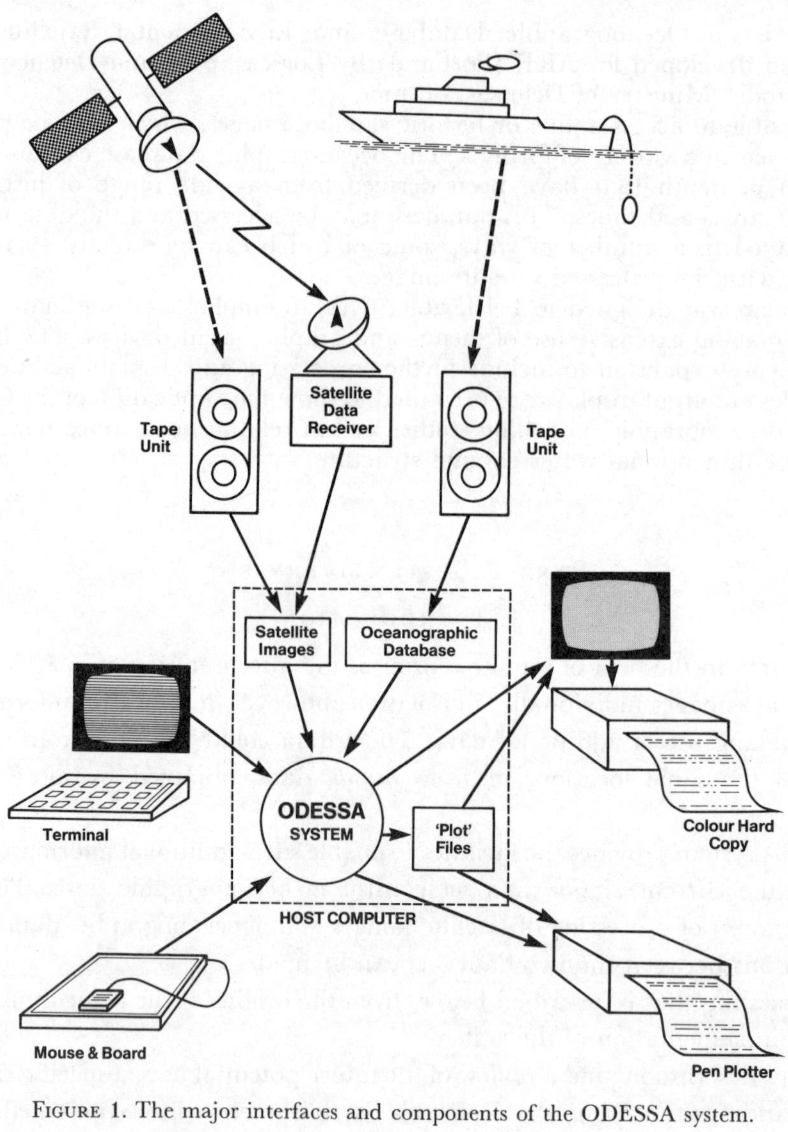

Figure 1. The major interfaces and components of the ODESSA system.

3. Requirements

The requirement was for the implementation of an environmental database, with the ability to relate information from the database to information obtained from remotely sensed infrared images. Figure 1 shows the major interfaces and components of the required system. It was required that the system should handle data from a variety of sources.

(i) Satellite images, obtained from the NOAA series of satellites, either on computer-compatible tape or directly (via a serial line) from a low-resolution satellite data receiver.

(ii) Oceanographic data in the form of temperature–salinity–sound-speed as functions of depth, at a large number of locations, covering a wide geographic area (the whole of the North Atlantic), and spanning a long period of time (many years).

(iii) User input via menus and by use of a graphics input device. This was in fact a mouse and board, which could also be used as a digitizer.

The output display formats were largely determined by the vector displays originally produced at SACLANTCEN, with the addition of further display facilities for raster images, and for the comparison of data sets from different sources.

4. Design considerations

The system was designed to be flexible and easy to use. Both of these are important attributes for a research tool, whose uses and users are likely to change with time.

Flexibility was promoted by the use of LUCID, an image processing software kernel (developed by Logica). With LUCID it is possible to construct complex sequences of image manipulations, which can be easily modified or extended. All the software is written in FORTRAN, and consideration was given to the need for a number of separate databases, each accessible by the same set of software, or a single large database split over several areas.

Ease of use was addressed by providing a menu-driven system, with three basic menu types: single choice, multiple choice, and a standard data input and edit menu. These input and edit menus provide default values for all parameters, which the user can change if required. Wherever possible, a graphics input device was used for specification of inputs, applying particularly to the determination of geographic limits on maps.

The special function keys, in addition to the menu applications, were also programmed to allow interactive analysis of the currently displayed image, at any stage of processing.

There were many other factors which were considered in the design of the system. For example, given the research environment within which the system was intended to operate, the level of automation was kept fairly low, allowing the user to control many of the operations interactively. In addition, the emphasis in the design was on flexibility rather than speed; this has allowed the inclusion of a larger number of features than would have been possible if speed of operation was the major consideration.

All these factors needed to be considered within the overall constraints imposed by the project timescales and budget. (There were six months available to produce the system, from initial design through to final delivery.)

5. System description

The hardware on which the system was based is summarized as follows:

DEC VAX computer, as the main host processor;

high-resolution colour graphics display, and ink-jet printer, for colour dumps of the screen;

large (up to A1 paper size) pen plotter, for hard copy of vector outputs from the system;

terminal and keyboard for user-control of the system;

mouse and board, for control of cursor on the graphics display, and for manual digitization of data;

satellite data receiver, connected to the host computer by a serial line.

Processing within the system can be broken down into five broad categories. Two involve only the oceanographic data, these being database loading and accessing, and database analysis options. One category is involved with just the satellite data: satellite image processing. The two data sets are related in the comparison facilities between images and database. Finally, within all of the categories are the special function key operations. The functions performed within each of these areas are now described in greater detail.

(a) *Database loading and accessing*

The system currently allows oceanographic data to be input in any one of three standard formats, from magnetic tape, or from disk file. Alternatively, individual data profiles can be input manually with a digitizer.

To optimize data-access times, the profiles are stored according to geographic areas, and instrument types. Once in the database, the data can be accessed in a variety of ways.

(i) By location, specified by the graphics input device to determine the limits of the search area on a map of the whole region, or numerically by specifying the minimum and maximum latitude and longitude, or the search area can be determined by the limits of a specified satellite image. The geographic location of the data can be further refined to a track of selectable width, a circular region, or a rectangular subregion of the current search area.

(ii) By time, including intervals covering time of day, date and year. The time and date of a selected satellite image can also be used as a basis for specifying the temporal limits.

(iii) By data source, including the specific type of instrument used, the particular ship from which the data were gathered, the general data type and the data quality.

Having identified the data required in this way, the next stage is to analyse the data.

(b) *Database analysis options*

There are five main database analysis options:

horizontal sections, consisting of plots of the geographic distribution of data;

vertical sections, showing the variation of a specified parameter along a particular track, plotted against depth;

profile plots, showing the data as they are stored in the database, as parameter against depth, from particular profiles;

temperature against salinity plots, with the distribution of the selected data points displayed;

extraction of inputs to acoustic propagation loss models.

The geographic distribution display allows for various different representations of the data, including scatter plots (with or without associated values), and contours of selectable interval, width and colour. The data displayed can be temperature, salinity, soundspeed or depth at a fixed level of any other of these parameters. (Thus one can display, for example, isotherms at a particular depth, or the depths of a particular isotherm.) To assist in the location of the data, they can be represented on a choice of map projections, with coastlines, gridlines, and detailed chart borders.

The vertical section display has the same choice of data representation options as the horizontal section. In addition, the sea-bed topography can be simultaneously displayed.

The graphical representations have the option of automatic scaling of axes to fit the available data, or the axis limits can be user-specified.

Likewise, the inputs to the propagation-loss models can be specified in detail by the user, or typical default values can be used.

(c) *Satellite-image processing*

There are several stages in the processing of satellite images before they can be used for comparison with the oceanographic data extracted from the database. These stages are summarized below.

(i) *Image extraction*

This involves reading the image data from magnetic tape, or over a serial line, separating the image from the calibration data stored with each image line, and saving the multiband image on disk.

(ii) *Atmospheric correction*

This involves converting the satellite-observed radiance at different frequencies to observed temperatures, by using on-board blackbody calibration data. This conversion is performed by the algorithms recommended by NOAA (1979). The observed temperatures are then converted to estimated sea-surface temperatures by making allowance for atmospheric absorption and attenuation. There is now a choice of a dozen different atmospheric correction algorithms, each relying on the linear combination of observed temperatures in different frequency bands. These algorithms were derived from a number of sources (e.g. Bernstein 1982).

In addition, new algorithms can be easily introduced, if required. From this point onwards, the satellite image is stored as a 'temperature image', with pixel values calibrated in tenths of kelvins.

(iii) *Geometric correction*

The space view of earth needs to be converted to a standard map projection to allow direct comparison with the oceanographic data. This is achieved by modelling the orbit of the satellite, and the Earth–satellite geometry. To produce a useful model, the user is required to specify various orbital parameters, and to specify the required location, size, scale and projection of the corrected image. The limits of the corrected image are then displayed on the uncorrected image as an appropriately distorted polygon. To obtain accurate location of the transformed image, the coastline overlay can be used as a 'ground control point', by aligning the overlay with those visible portions of coastline on the image.

The image is now in a state where comparisons with oceanographic data can be made.

However, there are various further enhancements which can be applied to the satellite image, if required. These include:

identification of cloud, creation of a cloud overlay, and limited interpolation beneath clouds (to provide an estimate of sea-surface temperatures under cloud-affected areas); further work is still required in this area, however, useful results can already be obtained (Gower 1985);

coastline, landmass, bathymetry overlays;

the ability to combine cloud and land overlays to give a composite image with land represented black, cloud white, and the visible sea-surface various shades of grey or coloured according to a user-controlled temperature scale;

a range of image-processing facilities such as median filtering, edge enhancement and contouring;

extraction and display of observed sea-surface temperature (SST) along a selected track.

(d) *Comparison facilities between images and database*

The main comparison facility available is the ability to display satellite images to the same scale and projection, and covering the same geographic areas as the output from the database analysis software. Figure 2 provides a representation of this facility, which is a key component of the ODESSA system.

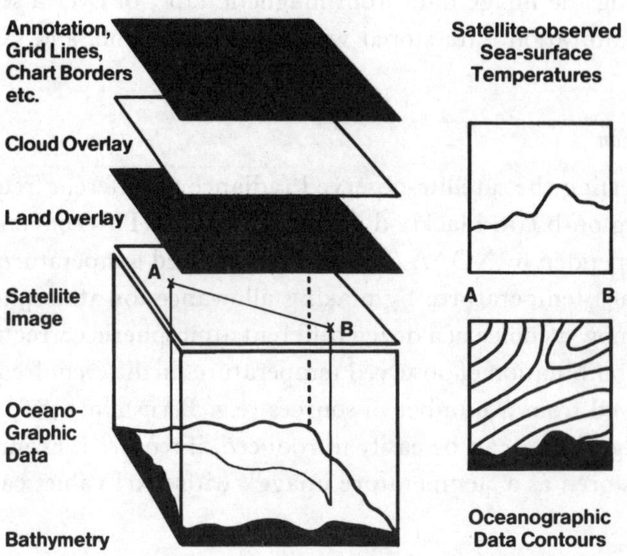

FIGURE 2. Relations between the various data sets and displays in the ODESSA system.

It is also useful to be able to display data from different sources in the same format, so the system allows the oceanographic data to be displayed as a raster image, with the same colour coding as the satellite image.

Alternatively, the satellite image can be contoured, and related to the contour output from the database analysis suite.

In addition, it is possible to show the limits of satellite images overlapping the current search area, and the overlapping portions of those images.

Any vector output can be put on the overlay plane, and superimposed on any raster image.

Furthermore, any vector output can be saved in a device-independent 'plot file' for later display via the colour monitor, pen-plotter, or graphics terminal.

The screen can be split into four sections, and separate displays put in each quarter, to allow for direct comparisons on a single screen.

As a further aid to interpretation of the images, the bathymetry of the specified region can be displayed as an image, with depth scale. Alternatively depth contours can be produced, and overlaid, if required.

The sea-depth information obtained from the bathymetry data is combined with the sound-speed profiles derived from the database for use in the input to the propagation-loss models. The track selected for display of a vertical section (along track against depth) can also be used directly, to extract the sea-surface temperature from a satallite image. Statistical analysis of direct comparisons between database-derived information and satellite-image-derived data can be made, and graphical representation of these can be displayed.

(*e*) *Special function key operations*

Operations using special function keys are available at almost any stage of processing within the ODESSA system. They allow interactive display manipulation, incluing stretching grey-scales and setting false colours, pan and zoom, screen annotation, and screen deletion.

In addition, when a temperature image is displayed, the temperature at any point in the image can be obtained and a temperature scale added.

Likewise, when a bathymetry image is displayed, depth at any point and a depth scale can be obtained.

6. Future developments

There are many ways in which this system could be further developed; these include the following.

(*a*) Extension to include other satellites and data types. Also the data need not be restricted to oceanographic data; any spatially dependent information could be used.

(*b*) Closer integration of data derived from different sources, for example:

integration of images obtained from different satellites;

use of *in situ* sea-surface temperatures to assist in the interpolation of satellite images beneath cloud;

performance of some form of 'interpolation' between data sets; this includes, for example, temporal 'interpolation' of the same geographic area, to derive information about the surface motion (see, for example, Clark *et al.* 1981; Ninnis *et al.* 1986), and spatial 'interpolation' between data sets at different depths.

In the long term an approach such as this could lead to the ability to produce predictions of the sea temperature at depth, based on sequences of cloud-affected satellite images of the sea surface and historical *in situ* data below the surface.

(*c*) Develop the user interface further, in particular in the areas of locating ground control points, and of cloud identification, where knowledge-based systems could have a useful role.

Part II. Application

A preliminary examination of data gathered during a survey of the Iceland–Faeroes region with the ODESSA system.

7. Data description

The application of satellite remote sensing to oceanography is of great importance to the Ocean Science Division at the ARE, U.K., with particular emphasis on the task of applying remote-sensing techniques in support of acoustic propagation studies. ODESSA was used in an investigation of the main oceanographic features of the Iceland–Faeroes region, in particular, the Iceland–Faeroes front, which forms the northeastern boundary of the Atlantic Ocean. It is created by the meeting of the cold waters of the Norwegian Sea in the north with the warmer, more saline waters of the North Atlantic Ocean. The frontal position is fixed by the underlying bathymetry where a broad ridge of less than 500 m depth stretches from the Icelandic Continental shelf to the Faeroe Islands.

For many years this region has been the subject of a number of investigations (see, for example, Dietrich 1967), following the discovery of an overflow of cold Norwegian water that spills over the ridge and spreads south to form part of the Atlantic Deep Water Body (Worthington 1976). Figure 3 shows the characteristics of the front with its location in the upper 25 m. The front extends from approximately 64° N at an angle of about 40° with the Iceland–Faeroe ridge, its apex being on the Icelandic continental shelf. In the region of

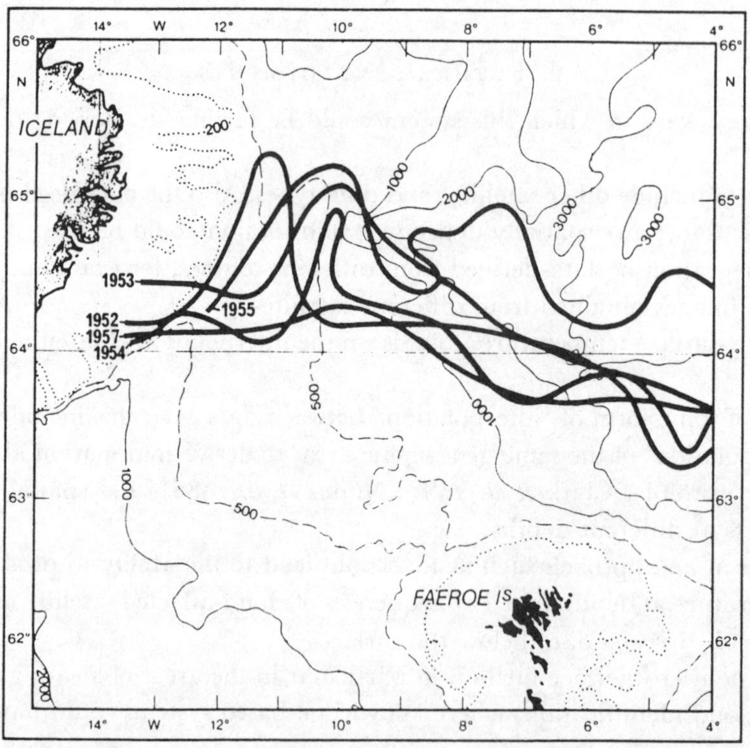

Figure 3. Locations of the Iceland–Faeroes front in the surface layer (0–25 m) in May and June of the years 1952 to 1957. The lines correspond to the 35.0 p.p.t. (by mass) isohaline, which represents the zone of maximum horizontal gradients (compiled from *Annls Biol* (1952–1957)).

100–150 km along the front from the Icelandic coastline, at the sea surface, lies an intrusion of warm Atlantic water covering a large area of the sea surface extending northwards into the colder Norwegian Sea water. The frontal area between the intrusion and the Icelandic coast appears to be dominated by eddying and meanders, while the area to the east of the intrusion is ill-defined and further displaced from the Iceland–Faroes ridge, which could be due to the greater amount of mixing in that region or the general increase in depth. Either of these, combined or in part, could be a factor influencing the general trend of the front.

In situ sea truth data and infrared satellite images were gathered during a survey of this region from 8–28 June 1986, and provide corroboratory evidence for the frontal position as shown by the satellite images, with the combination of the frontal position from the satellite image and the subsurface hydrography allowing the development of the concept of the front as a three-dimensional feature. The preliminary examination of survey data has made extensive use of the facilities available in the ODESSA system. In particular, the facilities within ODESSA have been used to analyse:

the spatial and temporal variability of the Iceland–Faeroes Front;

the effect of bathymetry on the Front;

temperature differences between remotely sensed and sea truth data.

The distribution of survey measurements is shown in figure 4; the measurements consist of 382 expendable bathythermograph temperature profiles. Several cloud-free and partly cloud-free advanced very-high-resolution radiometer images from the *NOAA*-9 satellite were available for the period. The image for 21 June 1986 is displayed in figure 5. A selection of representative temperature against depth profiles are shown in figure 6. In addition, historical data, available within the system, were used to compare with features observed from the survey data and highlighted previously by other workers in this field.

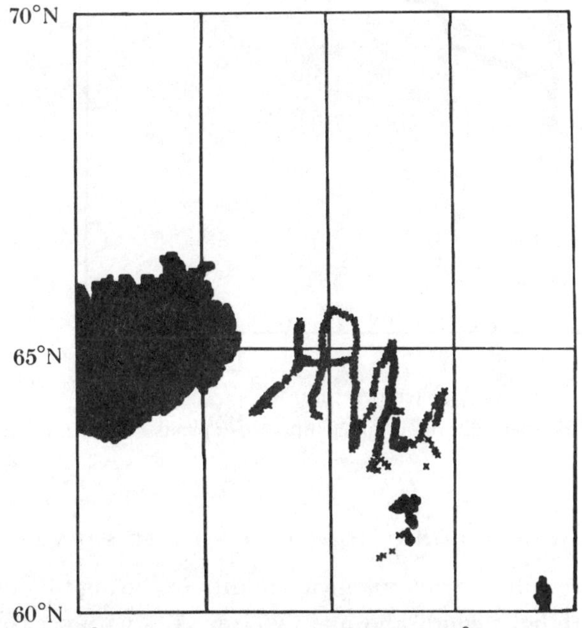

FIGURE 4. Scatter plot of expendable bathythermograph profiles gathered during a survey of the Iceland–Faeroes region in June 1986.

FIGURE 5. Geographical location of a satellite subimage for 13 h 34 GMT 21 June 1986, with the upper-left and lower-right corners impinging on the coastlines of Iceland and the Faeroes respectively.

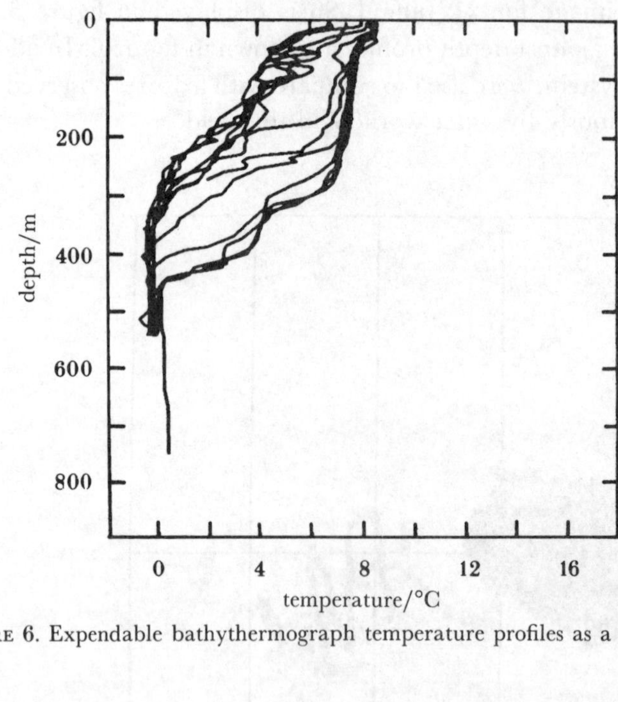

FIGURE 6. Expendable bathythermograph temperature profiles as a function of depth.

8. METEOROLOGICAL CONDITIONS DURING THE SURVEY PERIOD

An examination of the prevailing meteorological conditions during the survey period showed an anticyclonic pattern with light winds and mainly clear skies with occasional fog. However, a period at the beginning of June and in the middle of the survey period produced cyclonic conditions with deep depressions forming to the west of Iceland producing fresh to strong southwesterly winds in the Iceland–Faeroes region.

9. Satellite image analysis

All cloud-free satellite images for June 1986 were incorporated into the ODESSA satellite image directory. The subimage for 21 June (figure 7), presented in Mercator projection, shows a sea-surface-temperature representation of the area, with the warm Atlantic water dominating the central upper part of the image and the colder Norwegian Sea water to the north. The image has been processed and enhanced to show the frontal area with associated eddies and

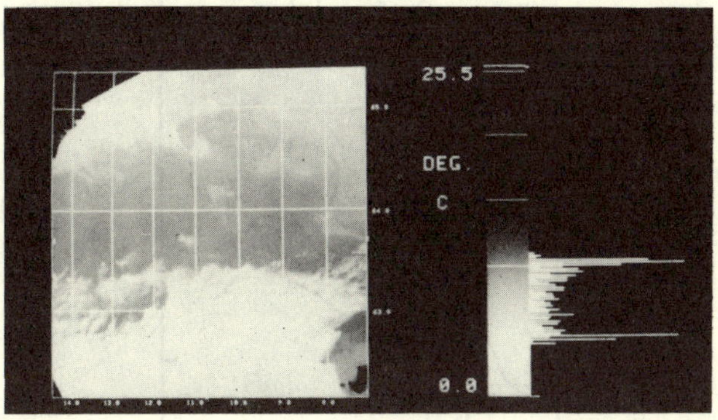

Figure 7. *NOAA*-9 high-resolution subimage, obtained from Dundee University, atmospherically and geographically corrected and shown here in Mercator projection as a 'temperature' image; i.e. each pixel represents absolute values of sea-surface temperature as shown on scale. Image, T892W; location, S.E. Iceland front; date, 21.06.86; time, 13h34; satellite, N9.

a well pronounced cyclonic hook protruding from the western side of the warm intrusion evident in the upper central part of the subimage. Other satellite images for the period portray the hook in various stages of decay and development suggesting a fairly brief life cycle, though the lack of images in sequence prevent conclusions being drawn regarding the period of this feature.

Another interesting feature is the cold eddy in the centre of the subimage at approximately 64.3° N, 10.5° W. Earlier images seem to indicate that the formation of the eddy may be due to a cold tongue of Norwegian Sea water intruding across the unstable frontal interface into the warmer Atlantic water and subsequently being cut off by the warm water of the cyclonic hook feature overriding the cold dense water of the intrusion. An investigation of this feature with *in situ* data corresponding to a track through the eddy, indicates a modified surface temperature structure with stronger temperature gradients at depth. The structure and behaviour of the front and associated mesoscale features have been the subject of many studies (see, for example, Willebrand & Meincke 1980; Smart 1984).

Meincke (1975) indicated that the eddy generation might be linked to changes in the atmospheric pressure distribution. However, Willebrand & Meincke (1980) indicated that baroclinic instability and the tendency of the mean bottom currents to follow the isobaths, causing eddy-like structures to emanate from the near-bottom layer, were the likely sources of eddy generation and intermittent overflow and only occasionally, during the passage of strong atmospheric cyclones, would forcing be likely to play a dominant role.

10. Temperature comparisons

The relation between the direct measurement of sea surface temperature by shipborne instruments and the infrared-derived measurements of the top millimetre skin layer of the sea–air interface has been tentatively assessed by comparing data obtained along the ship's track with corresponding pixel data on the image. Figure 8 is an example of a case in which

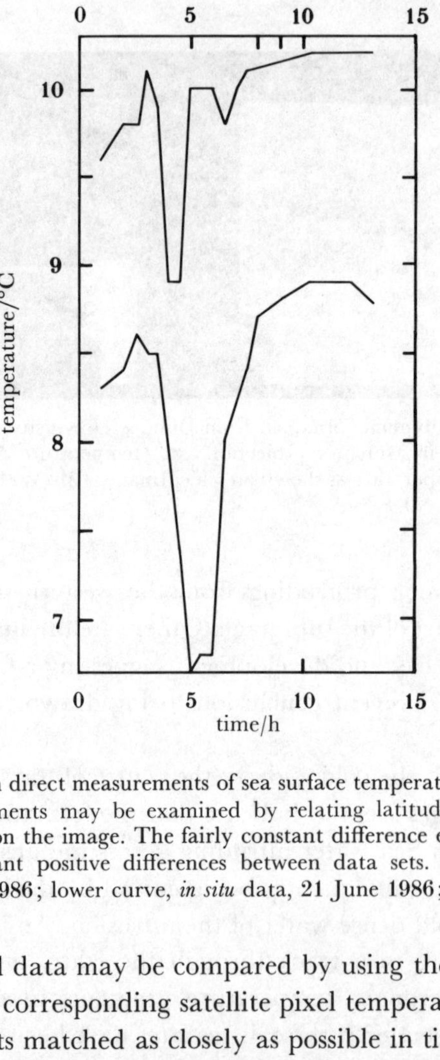

Figure 8. The relation between direct measurements of sea surface temperature by shipborne instruments and the infrared-derived measurements may be examined by relating latitude and longitude of *in situ* data with corresponding pixel data on the image. The fairly constant difference evident here may well point to strong solar heating with resultant positive differences between data sets. Upper curve, satellite-derived data, 21h13 min 55 GMT, June 1986; lower curve, *in situ* data, 21 June 1986; RMS difference, 1.86 K.

in situ and satellite-derived data may be compared by using the ODESSA system. The *in-situ* temperature data and the corresponding satellite pixel temperature values are plotted against time with the two data sets matched as closely as possible in time, to enable a comparison to be made. The RMS difference between the two sets is 1.86 K; the general trend of the temperature distribution is clearly mirrored in both data sets.

More important to this qualitative account is the relation between the direct measurements of sea surface temperature and the temperature related features depicted by the image. An appraisal of the data presented in two forms: vector and raster mode, enabled comparisons to be made of the contoured data from the satellite image and the raster image generated from the contoured *in situ* data (figure 10). A comparison of the overall grey level distribution in both images reveals the lighter shades in the *in situ* image and, correspondingly, the contoured *in situ* data indicate a lower temperature field with less complex and smoother defined contours than those in the satellite-derived contours. Generally, however, comparison of the *in situ* data with those derived from the satellite shows similar sized temperature ranges for the water masses and frontal water with the subimage appearing, as in figure 8, 1–2 K warmer. Moreover, meteorological conditions for the period covered by the satellite image were favourable for strong solar heating to produce conditions of positive differences to occur between satellite-derived and *in situ* sea surface temperature, indicating a possible explanation for the warmer 'skin' temperature.

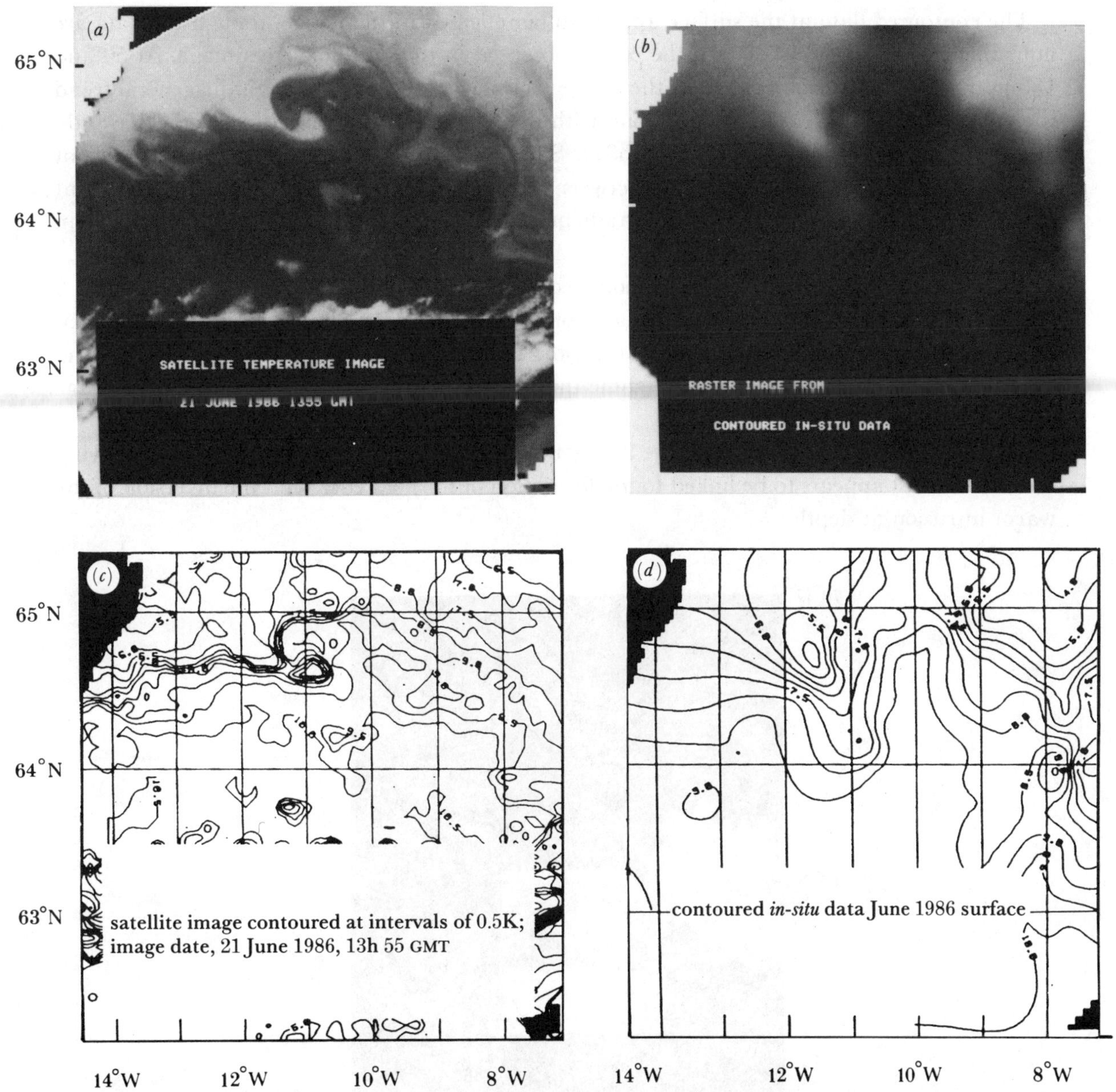

FIGURE 10. Satellite-derived and *in situ* measurements of sea-surface temperature presented in raster and vector mode. Both sets of contoured (vector) data show the general distribution of warm Atlantic water with the cooler Norwegian Sea water to the north. Comparison between overall grey level distribution in both raster images indicate the lighter shades in the *in situ* image corresponding to a lower temperature field.

11. Intercomparison between data sets

The satellite subimage, overlaid with *in situ* contoured temperature data, provides information on the vertical structure of the ocean in this region. Figure 9, plate 1, shows data gathered during the period 8–28 June 1986 representing the data set to correspond with the 21 June image. An initial study of the satellite images available for this period indicated no significant change in the location and general shape of the frontal feature. The combining of *in situ* data over this timescale enabled the approximate location and shape of the front to be determined, allowing study of the underlying water column.

The contoured data at the surface appear to be reflected in the image, though the data are not sufficiently resolved to discern the eddies observed in the middle of the image. At 50 and 180 m the warm intrusion evident in the image is clearly discernable, though the general trend is a southward incline of the frontal zone with depth. This spatial variation of the Iceland–Faeroes front was studied by Smart (1984). Smart concludes that the location of the front can be displaced by as much as 150 km between 50 and 350 m depth. The overall displacement of the zone of maximum temperature gradient observed in figure 9 is not as apparent as in the observations of Smart.

An interesting feature is the well pronounced meandering of the front, which presents an S shape at 180 through 350 m depth. Smart commented on this S shape, observed during his appraisal of the frontal displacement at depth, though his observation was of a recurring inverted S-shaped meander closely matching the 500 m depth contour at the southwestern edge of the Iceland–Faeroes ridge.

Though the feature evident in figure 9 may be topographically driven, the strong recurving to the northeast appears to be linked to the formation of the cold eddy and the inversion of the warm intrusion at depth.

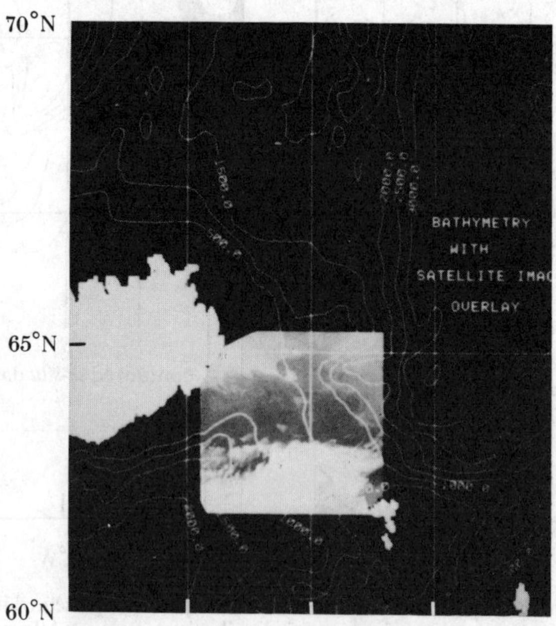

Figure 11. Satellite subimage overlaid on bathymetry contours. The spacing of the contours may be selected interactively by the user.

FIGURE 9. Contoured *in-situ* temperature data at significant depths of near-surface, 50m, 180m and 350m depth overlaid on the satellite subimage; an example of how *in-situ* data can be compared with accurately registered and atmospherically corrected satellite images.

Figure 11 shows the bathymetric contours in the area covered by the satellite subimage. The correlation between the surface and the bottom structure is small though the warm intrusion follows the northward extent of the Iceland–Faeroes ridge. However, the zone of maximum temperature gradient evident in all the contoured data in figure 9 at approximately 64.5° N, 12.0° W correlates well with the bathymetric contours associated with the South Iceland Channel. Moreover, the frontal feature appears to follow the 500 m contour at 350 m depth, an observation also made by Smart, though the northeastern edge of the front appears to follow the northern boundary of the Iceland–Faeroes ridge.

12. Vertical section through the frontal zone

A section was taken through the frontal zone corresponding to a track of the survey vessel. Figure 12 shows temperature contours generated from the *in situ* data along the ship's track and exposes the slant of the front showing clearly the outflow of cold Norwegian Sea water spilling

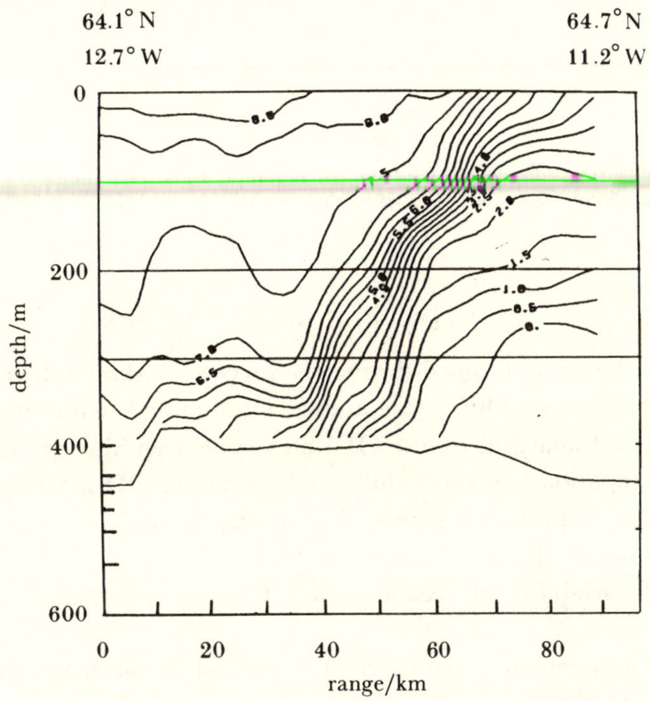

Figure 12. Temperature section along the crest of the Iceland–Faeroes ridge, obtained by the survey vessel in the period 22–23 June 1986.

over the ridge with the warm Atlantic water to the south. The surface expression of the frontal zone correlates well with the satellite image indicating the zone of maximum temperature gradient. Additionally, the ODESSA system allows the creation of data files for input into propagation loss models. Figure 13 shows selected sound velocity profiles produced from the track temperature data (figure 12) and subsequent output, showing the effect of the oceanography on the propagation of sound with a large transmission loss through the frontal zone.

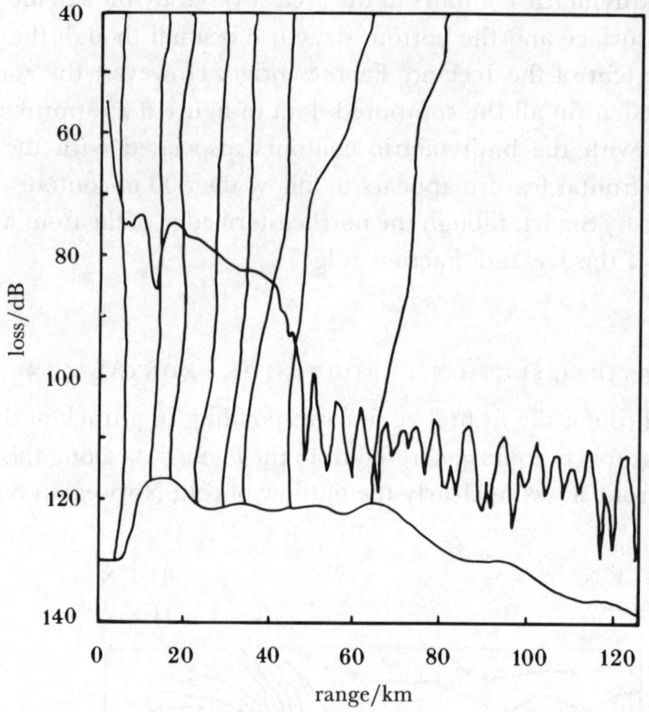

FIGURE 13. Series of sound velocity profiles generated from track. temperature data (figure 12) and output from a parabolic equation propagation loss model showing transmission loss as a function of range, with bottom topography superimposed.

13. Conclusions

The initial analysis of the data obtained during the survey has shown the relative ease with which large amounts of oceanographic *in situ* and corresponding satellite image data can be processed, manipulated and analysed by the ODESSA system within very short timescales.

The preliminary investigation has also highlighted the use of ODESSA for the following.

(*a*) Oceanographic area appraisals for pretrial planning exercises.

(*b*) Oceanographic analysis of:

spatial and temporal variability of oceanographic features (temperature, salinity, sound speed);

comparison of *in situ* and remotely sensed data as an aid in the development of satellite atmospheric correction algorithms;

(*c*) deriving water-column structure from satellite infrared images and *in situ* data as an aid in oceanographic modelling and acoustic propagation studies.

In conclusion, the ODESSA system represents a step in the future direction of the application of remotely sensed data, in that it involves integration with data from other sources. It is a research tool which has considerable potential for further development in the integration and analysis of remotely sensed images with *in situ* data.

References

Bernstein, R. L. 1982 Sea surface temperature estimation using the *NOAA*-6 satellite AVHRR. *J. geophys. Res.* **87**, 9455–9465.

Clark, J. R. & LaViolette, P. E. 1981 Detecting the movement of oceanic fronts using registered *Tiros*-N imagery. *Geophys. Res. Lett.* **8** (3), 229–232.

Dietrich, G. 1967 In *Review of the international 'overflow' expedition (ICES) of the Iceland–Faeroe Ridge, May–June 1960* (CPI Pour L'Exploration de la Mer. Rapports et Proces verbaux, vol 157), pp. 268–274.

Gower, J. F. 1985 Reduction of the Effect of Clouds. *J. Remote Sensing* **6** (8), 1419–1434.

Meincke, J. 1975 Overflow 73 – evidence for atmospheric forcing of Arctic water events. ICES Hydrogr. Comm. CM 1975/C: 29.

Ninnis, R. M., Emery, W. J. & Collins, M. J. 1986 Automated extraction of pack ice motion from AVHRR imagery. *J. geophys. Res.* **91** (C9), 10725–10734.

NOAA 1979 Data Extraction and Calibration of TIROS-N/NOAA Radiometers. Technical memorandum NESS 107. Washington, D.C.: Department of Commerce.

Robinson, I. S. 1985 *Satellite oceanography*. Chichester: Ellis Horwood Ltd.

Smart, J. H. 1984 Spatial variability of the major frontal systems in the North Atlantic–Norwegian Sea, 1980–81. *J. phys. Oceanogr.* **14**, 185–197.

Willebrand, J. & Meincke, J. 1980 Statistical analysis of fluctuations in the Iceland–Scotland frontal zone. *Deep Sea Res.* **27A**, 1047–1066.

Worthington, L. V. 1976 *On the North Atlantic Circulation*. Baltimore and London: The Johns Hopkins University Press.

Phil. Trans. R. Soc. Lond. A **324**, 365–372 (1988)
Printed in Great Britain

Applications of remote image sensing with the NASA massively parallel processor

By M. Halem

Space Data and Computing Division, NASA/Goddard Space Flight Center, Greenbelt, Maryland 20771, *U.S.A.*

[Plates 1–4]

Satellite observing systems are producing image observations of the Earth's surface and atmosphere with spectral and spatial resolutions that result in data rates that current general-purpose computing systems are incapable of processing and analysing. As a result, current processing systems have been able to analyse only limited amounts of image data with less than optimal algorithms for generating high-quality geophysical parameters. A massively parallel processor (MPP) is operationally available at NASA/GSFC for routine image-analysis applications. Research studies with the MPP are being pursued in the area of interactive spatial contextual classifications for the land thematic mapper data, automatic SIR-B stereo terrain mapping, ice-motion detection, faint-object image restoration and other general purpose ocean and land image-processing systems. Several applications are presented comparing the MPP products with enhancements of imaging data with standard image-processing methods. Finally, a work-station parallel processor for space station on-board image processing will be described.

1. Introduction

Satellite observing systems are capable today of spectrally imaging the Earth's surface and atmosphere at data rates far exceeding our ability to process or extract their full information content in any practical manner. This has resulted in limited duty-cycle use of our instruments with high data rates and, very often, archiving vast amounts of unprocessed data for possible future access. With advances in charge-coupled device (CCD) technology, more of the passive sensors with lower data rates will evolve into imaging sensors with even higher spectral and spatial resolutions provided by their densely packed arrays of detectors. As we move into the permanently manned space station era, an Earth-observing system is planned to carry these facility scale instruments so as to provide continuous images that will monitor the long-term changes in the Earth's bio-physical and chemical cycles. Along with developments in passive sensing, massive image data sets will also be produced from space-borne active sensors that use very large real or synthetic apertures. The impact of these observing systems could lead to a revolution in our understanding of Earth system science, but to achieve that goal a radically new technology based on parallel computing architectures and algorithms will have to be developed.

Recognizing the enormous image processing requirements for the Earth-science sensors and similar imaging requirements emerging in the planetary, solar-terrestrial and astrophysical sciences, NASA embarked, in 1979, on a pilot project to design and build a massively parallel processor (Potter 1985) that will begin addressing these problems. The system was delivered in 1983 by the Goodyear Aerospace Corporation (Batcher 1980) and shortly thereafter was made available on a selective basis to a wider community. A research programme was initiated in 1985, through a national solicitation, that established an MPP working group that is exploring the algorithmic potential of this uniquely powerful architecture. In this report, I present selected applications from the MPP working group investigations that dealt with remote Earth-sensing image algorithms. Section 2 contains a short description of the MPP system, and §3 describes automated Shuttle Imaging Radar-B (SIR-B) stereo terrain mapping and the application of similar techniques for sea-ice motion detection. Section 4 presents multispectral classification algorithms applied to *Landsat* thematic mapper data. Section 5 outlines potential future developments and applications of massively parallel architectures that could be applied to on-board space-station processing.

2. Description of MPP

The MPP is an integrated system of 16384 processors, each containing its own local memory of 1024 bits, and a global data staging memory of 32 megabytes. All processors perform the same single-bit arithmetic and logical operations at instruction speeds of 100 ns (10 MHz). The processors are arranged in a grid connected array 128×128 with single-bit transfer to nearest neighbours in north, south, east or west directions simultaneously executed at the same cycle speeds. Automatic east–west and north–south wrap-around symmetry occurs also at the same clock cycle.

Transfer from the global data staging memory to the array memory is at 80 megabytes per second and consists of 128 bits moving into or out of a column or row simultaneously in two cycles. Three controllers govern the execution and synchronization, and global broadcast of instructions to all processors. The system is accessed through a front-end Digital Equipment Corporation VAX 11/780 processor and a DR-780 interface to the MPP with a transfer rate of 5 megabytes per second. A functional sketch of the hardware layout is shown in Fischer (1987). A high-level parallel extension of the PASCAL language is available for general scientific programming on the MPP. Assembly coded macroinstructions for full floating point operations are transparently provided under MPP PASCAL. Programs are compiled and loaded through the VAX front-end.

Arithmetic performance speeds for addition and multiplication can range from 40×10^9 operations per second for 1-bit addition to 1.68×10^9 operations per second for 32-bit addition and approximately a factor of ten slower for fixed point multiplication. Floating point 32-bit addition and multiplication operations are performed at 450 and 215 MFLOPS (million floating-point operations per second), respectively. Overall, MPP machine computing performance relative to a CYBER 205 and Cray-1 for vector modelling computations is three to five times faster, respectively, and upwards of 100–1000 times faster for appropriate fixed-point algorithms.

3. Automated stereo elevation mapping from space

In this section, we present results of a fully automated stereo matching algorithm for the determination of terrain elevation data from space-borne synthetic aperture radar (SAR) imagery data. This is work that was developed at Goddard Space Flight Center and implemented on the MPP (Strong & Ramapriyan 1987). Their MPP algorithm is based on a low-level vision technique (Marr & Poggio 1979 as extended by Quam 1984).

The algorithm was applied to a pair of overlapping SIR-B images acquired during the October 1984 Space Shuttle Challenger flight. The images were taken over a plateau region of Bangladesh at incidence angles of 25° and 42°, respectively. As a result of the viewing geometry, a linear scale change in the slant range (stereopsis) direction is first applied to the latter (i.e. test) image to produce the same pixel resolution as the former (i.e. reference) image. For each pixel in the reference image, a rectangular neighbourhood surrounding it is correlated with a set of neighbourhoods corresponding to each point in a search area in the test image. The disparity function is then determined by associating in the stereopsis direction the location of the corresponding centre pixel of the neighbourhood within the search area with the maximum correlation value. Badly matching regions of the disparity function are defined by adjacent pixels with jumps greater than one and filled in by an interpolation procedure based on an analogue to a finite-difference heat-flow equation applied to the region. The disparity function is then used as a geometric correction function to warp the test image so that it more closely matches the reference image. The algorithm is iterated for the entire image with successively smaller neighbourhoods. Figure 1, plate 1, shows the reference and test images overlayed after the initial linear warping and edge alignment. The two images are displayed with red and green, respectively, as maximum brightness values. One readily observes a broad plateau region with a surrounding river valley on the right. Figure 2, plate 1, shows the results of the matching algorithm after two iterations starting with rectangular neighbourhoods of 25×25 pixels and 13×13 pixels, respectively. The coincidence of the red and green in contrast with figure 1 indicates the agreement of the match. Figure 3*a*, plate 3, shows the disparity function with dark regions representing pixels in the reference image to the right of corresponding pixels in the test image and white regions to the left. Disparity is linearly proportional to elevation with dark areas corresponding to low elevations. By a three-dimensional graphics-rendering capability developed for the MPP (Dorband 1986), figure 3*b*, plate 3, displays a three-dimensional view of this region illuminated from a source approximately at the location of the shuttle at 42° from vertical. Overlayed on the disparity functions are the observed brightness values. An examination of a cartographic contour map of the same region available at NASA shows very good feature agreements between plateau, river, valleys and height gradients for the entire region. This calculation required more than 10^{11} operations in a little less than a minute which could not have been computed in less than a week of dedicated, around-the-clock VAX 11/780 CPU time. More importantly, these results establish the viability of generating reasonably accurate estimates of terrain height from space.

4. Sea-ice motion detection from time-varying SAR imagery

Unlike elevation mapping by stereo matching, where surface features are stationary in time and all disparities in multi-incident viewing image pairs are attributed to variations in elevation, detection of ice-floe motion assumes 'identical viewing geometries' in pairs of images and attributes local differences to motion within the scene. In this study, we report on algorithms for detection of ice motion developed for the MPP (Strong & Ramapriyan, 1987) and some preliminary results achieved thus far for two-dimensional ice motion detection.

The study is conducted with data acquired in 1978 from nearly identical viewing angles of the SAR sensor flown on *Seasat* as kindly provided by C. Elachi, of the Jet Propulsion Laboratory. We assume that a three-day interval as used in this study is sufficiently small that the dynamics of motion will not cause the distortion of the ice surface to the extent that an area in one image will not be recognizable in the other.

A major difference to be considered in extending the ideas of the automated stereo matching algorithm is the need to account for rotation and the fact that large changes are possible between image pairs. Because of the latter condition, we first apply a global cross correlation between the reference image and test image. This is performed efficiently by assuming periodicity in both horizontal and vertical directions and performing a two-dimensional fast Fourier transformation. Because translations of large coherent regions produce local maxima or peaks in the global cross correlation, the locations of these peaks can be used to define an approximate translation vector between the corresponding pixel of a given peak and the centre pixel of the global correlation. After translating the test images by an amount corresponding to a given local maximum, the local correlation technique of the type used for stereo matching is then used to identify the corresponding regions of points in the test image which match those in the reference image. A threshold value is used to identify those points in this region which will be stored in a two-dimensional array. The process is applied to all local peaks creating a sparsely filled, two-dimensional array representing the global displacement field. In accounting for rotation of large regions, we apply a sequence of small finite rotations to all points of the test image. We can fill the boundary elements of the test image if the original test image contains pixels outside the 512×512 region or else fill in by zeros. The process is then repeated for the rotated test image thereby building up the displacement functions. At the end of a complete rotation sequence, gaps in the displacement array are filled in, as in the case of stereo matching, by a finite-difference diffusion approximation. This algorithm is much more computer-intensive than stereo matching. At the time of this report, only the global correlation and a single translation have been completed. The investigators are in the process of assessing the size of regions which can be detected with the global correlation functions. Figure 4*a*, *b*, plate 3, shows two nearly identical views of Banks Island, located northeast of Alaska, taken by *Seasat* with a three-day interval. The local horizontal and vertical correlation values are calculated after performing a global cross correlation and translating the test image by an amount given by the correlation peak. In figure 5*a*, *b*, plate 4, the large solid dark regions represent small values of horizontal and vertical displacements as one expects after the test image has been translated. Figure 5*c*, plate 4, shows the maximum local correlation value at each pixel indicating a good match in the areas with a low amount of translation. The pixels in this region contributed to the local maximum in the global correlation function. The scheme does seem to be working as expected as far as we have thus far gone and holds extreme promise of being reasonably successful for automatic ice-motion detection.

5. Contextual classification algorithms

A major goal of multispectral remote sensing has long been the desire to segment and classify scenes of the Earth's surface according to their physical and biological spectral characteristics. The inherent difficulty of this problem lies principally in two sensor-related areas, namely, the non-uniqueness or even lack of wide band spectral signatures and the inadequate temporal and spatial resolution and sampling for any given scene. A third and often overlooked limitation has

Figure 1. Pseudo-stereo overlay of SIR-B shuttle images. Red and green indicate maximum radar reflection values, respectively.

Figure 2. Overlay after application of two iterations matching algorithm. The blending of red and green into yellow indicates good agreement resulting from warping.

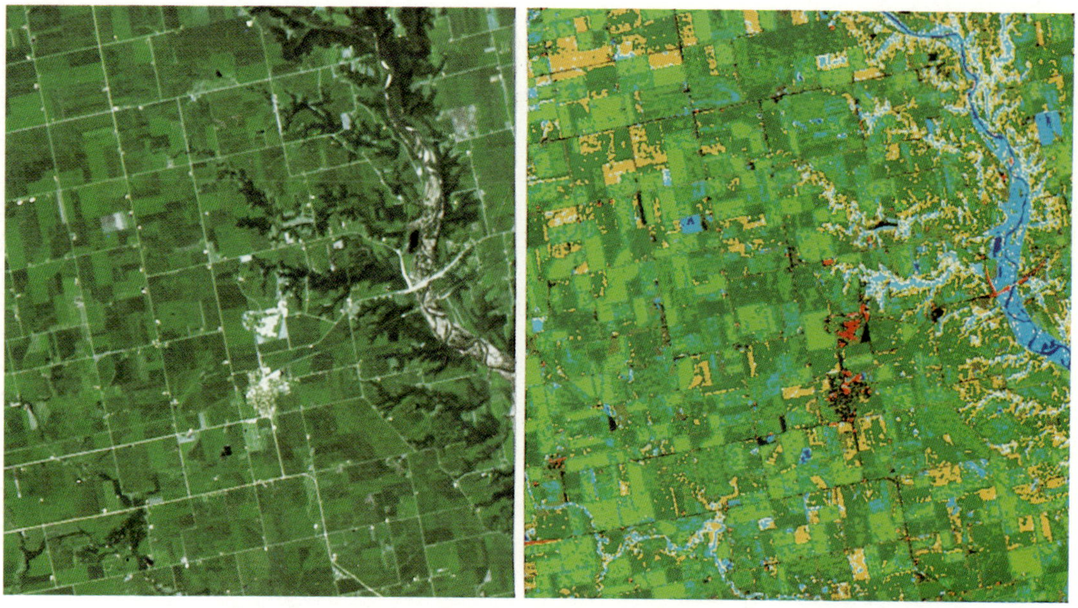

FIGURE 6. Portions of a *Landsat* scene before (left) and after (right) the application of 100 iterations of unsupervised ISODATA classification algorithm.

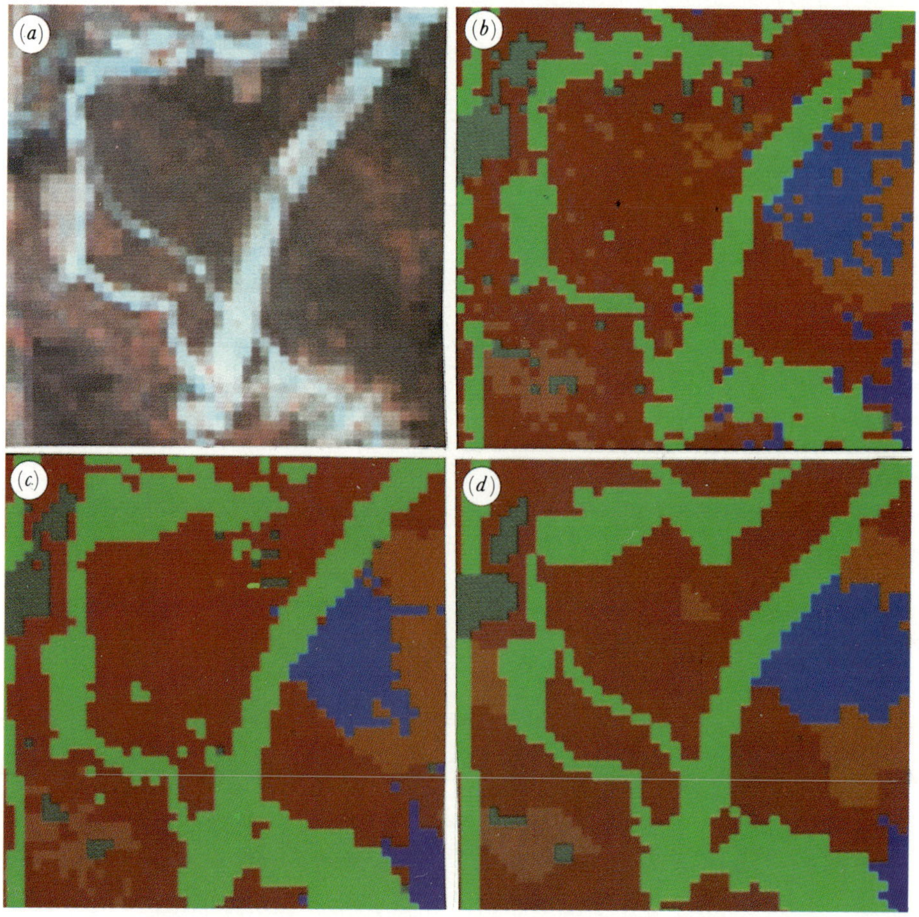

FIGURE 7. Three image representations of a *Landsat* scene: (*a*) unclassified; (*b*) after application of maximum-likelihood classification algorithm; (*c*) after application of five-point contextual extension of maximum-likelihood classifier; (*d*) ground-truth classification.

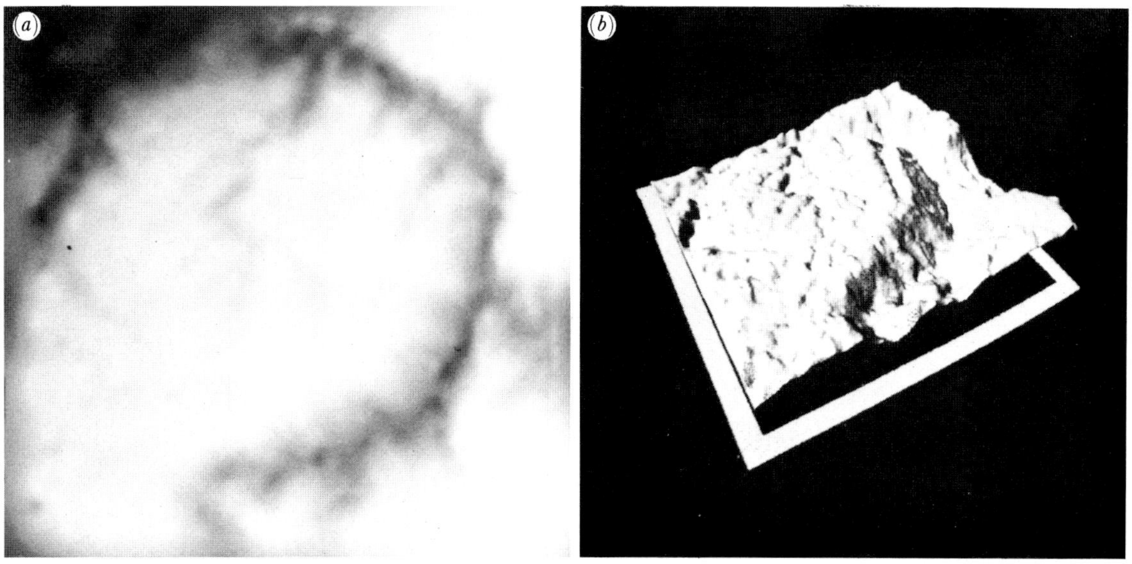

FIGURE 3. (*a*) *SIR-B* stereo disparity function after second iteration. Dark regions indicate low elevation and white regions indicate high elevation (*b*). Three-dimensional perspective rendering with stereo disparity function as height surface illuminated from a light source at an angle 42° from vertical.

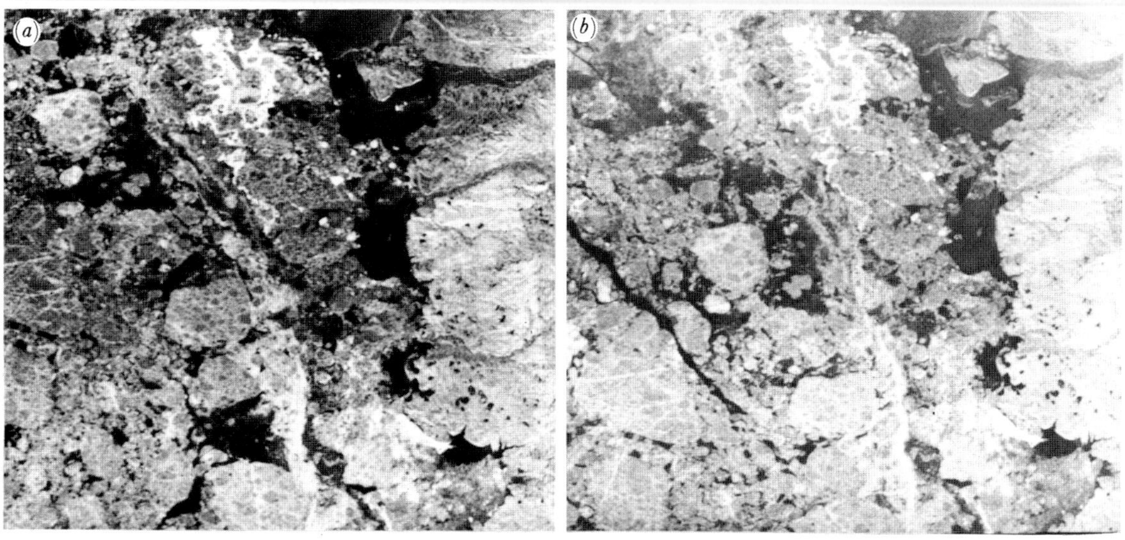

FIGURE 4. Near identical views of Bank Island, Alaska, taken by *SEASAT* synthetic aperture radar. (*a*) Image taken three days before image (*b*).

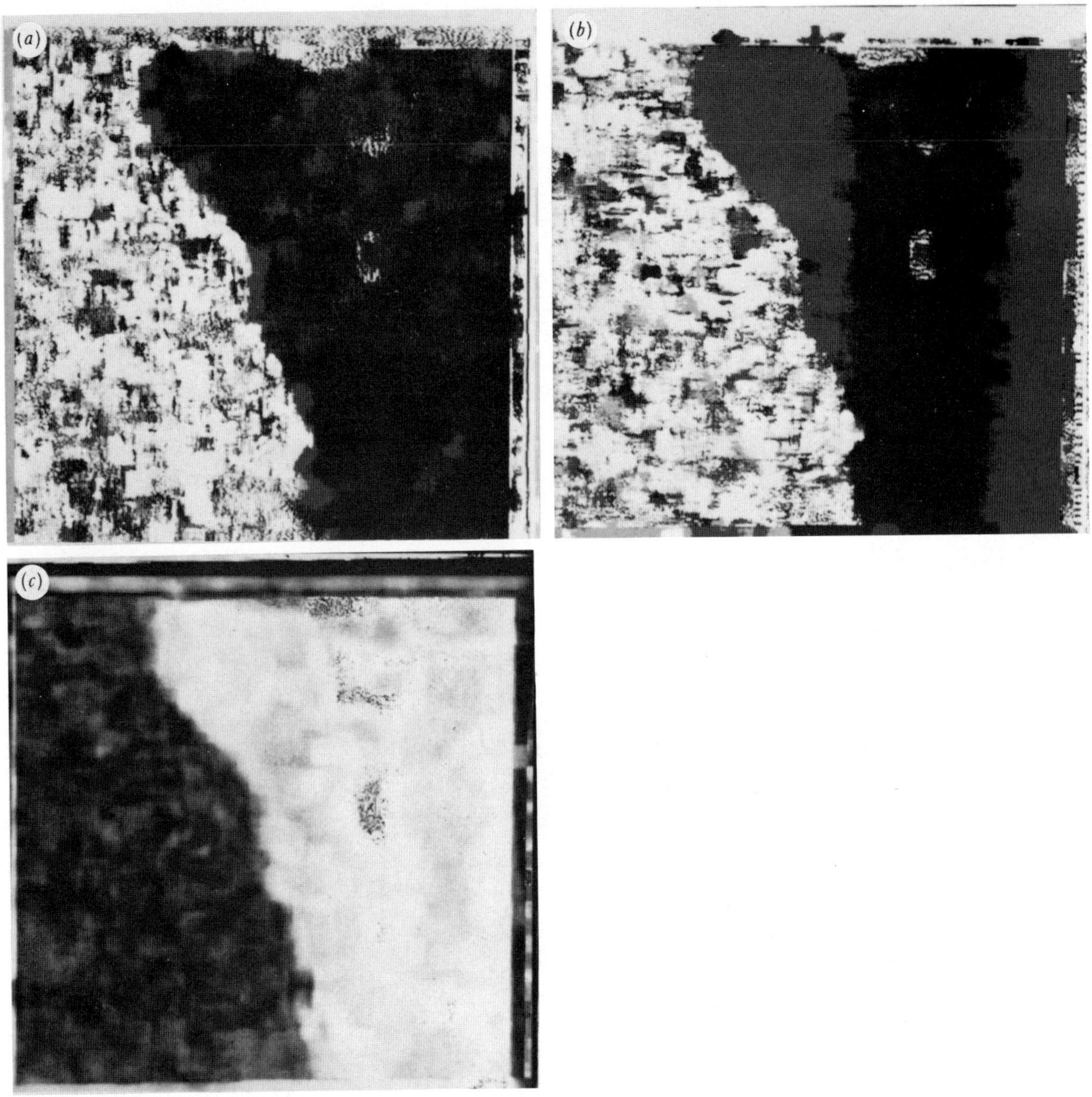

FIGURE 5. (*a*) Horizontal and (*b*) vertical translation functions, respectively. Large dark areas indicate little or no translation. (*c*) Match score values of displacement array. Brightness shows good matches within overlapping ice floes.

also been the inadequacy of current computational resources required for complex processing, analysis and classification of the massive volumes of pixel information available per scene.

Whereas both of the former factors are being addressed through the design of more advanced multispectral sensors with narrower spectral intervals, slower progress is being made in the development of improved image-analysis techniques. Although conceptual methods holding great promise have been available for some time, detailed testing progress has been limited until now because of the great demands on computational power that are needed to implement these techniques.

In this section, I report mainly on the use of the MPP to implement a contextual classification algorithm (Tilton & Strong 1984). Results are first presented for an unsupervised classification algorithm on the MPP which is needed as the initialization step for the generalized maximum likelihood classification algorithm. The unsupervised algorithm is known as ISODATA and is a straightforward spectral clustering method using the speed of the MPP to carry out the enormous number of calculations for a very large class of classification categories. Assume we are interested in M classification categories of a multispectral scene which has $N \times N$ pixels and where each pixel has L associated observed spectral bands. The algorithm consists of initially selecting M points in the spectral phase space of L-tuples labelled $m_i^{(0)}$, $i = 1, \dots, M$ considered to be the spectral centroids of all pixels in a certain category. The $m_i^{(0)}$ can be chosen by performing a principal eigenvector decomposition over all pixels in the scene and selecting the $m_i^{(0)}$ centroids along this principal direction. Each pixel in the scene p_k, $k = 1, \dots, N^2$ is associated with the centroid category $m_i^{(0)}$ if the condition that $\|p_k - m_i^{(0)}\| < \|p_k - m_j^{(0)}\|$ holds for all $j \neq i$. The norm $\| \;\|$ is taken as the usual euclidean L distance measure. The algorithm then consists of iterating for a new set $m_i^{(1)}$ obtained by calculating the mean for all pixels associated with $m_i^{(0)}$ and defining the new associated classes. Elaborations to incorporate cluster splitting, cluster combining and pixel weighting are ISODATA modifications planned, but not yet implemented, on the current MPP version. Figure 6, plate 2, illustrates the classifications starting from the image shown in figure 6*a* for a 512×512 *Landsat* TM scene over Iowa, after 100 iterations. The time to do this algorithm on the MPP was 18 s as compared with about 7 h on a VAX 11/780. This already demonstrates that the MPP can be used to convert ISODATA into an interactive tool for use by a scientist at a workstation remotely connected to the MPP.

Next, the maximum likelihood classifier (MLC) is applied to resolve the scene further according to *a priori* spectral class statistics obtained from training site data. Two sets of statistics were used in the study. One set characterizes the spectral information for each ground-cover class and is obtained from a per-pixel analysis of data from training sites within the TM scene being analysed. The training-site data had previously been correlated with field data giving the true ground-cover label for each pixel in each training site. The other set characterizes the contextual information for each pixel location and is compiled for individual pixels and their four nearest neighbours. The intuitive procedure whether doing MLC or generalized MLC algorithms consists in developing a decision rule for classifying each pixel. In the generalized MLC we applied the rule that associates the generalized pixel P_k with category i if $g_i(P_k) \geqslant g_j(P_k)$ for all $j \neq i$ and where $g_i(P_k)$ is given as a discriminant function for category i by $g_i(P_k) = \ln(D_i) - (P_k - M_i)^T C_i^{-1}(P_k - M_i)$, where P_k and M_i are single or spatially arranged vectors, C_i is the statistical covariance matrix associated with category i, D_i is the associated determinant of covariance matrix C_i.

The underlying rationale for embedding contextual spatial statistical information into the generalized maximum likelihood classifier rests on the assumption that *a priori* probabilities can be established for likely and unlikely geometric distributions of certain category types. For example, consider three classes of ground cover (forest, corn and water) represented in three different spatial configurations. Clearly, on a per-pixel basis, one would not expect to find a field of corn surrounded by water or a forest at a 30 m resolution. Although this example is simplistic, many situations do arise where we can incorporate probabilistic spatial distributions. Thus, we extend the associated phase-space vector tuple by considering as components of the vector the spectral band information of all four nearest neighbours in a prescribed order. This not only increases the pixel vector size linearly by the product of the number of spatial pixels and their bands, but also increases the complexity of the covariant matrix by the square of the increased vector size. For example, a five-point pixel stencil with seven bands increases the size of the covariant matrix from 7×7 to 35×35. For a *Landsat* TM scene this involves 6967×5969 pixel elements, well beyond the capability of any of today's mini processors.

Preliminary results have been obtained (Tilton 1986) on the application of this algorithm to an airborne multispectral scanner (ABMSS) scene acquired on 29 July, 1978, which covers a 2.8 km × 2.8 km area in British Columbia, Canada near the Anderson River. This data set is a part of a SAR–MSS data set that was acquired, preprocessed, and loaned to us by the Canada Centre for Remote Sensing (CCRS), Department of Energy, Mines, and Resources, of the Government of Canada. The data set covers a forested area consisting of areas of pure Douglas fir, Douglas fir mixed with lodgepole pine, Douglas fir mixed with cedar, Douglas fir mixed with hemlock, hemlock mixed with Douglas fir, hemlock mixed with cedar, forest clearings (including clearcuts and logging roads) and rock outcroppings. Figure 7*a*, *b*, *c*, *d*, plate 2, shows the unclassified image, a maximum-likelihood classification assuming each class has equal *a priori* probabilities, a five-point contextual maximum-likelihood classification, and the ground-truth classification, respectively. Figure 7*c* clearly shows much improved spatial continuity within the classification categories. Performing a statistical comparison with ground truth for the entire 2401 pixels which were classified, the spatial MLC had a 79.4% correct classification compared with 75.5% for the standard MLC for the eight classes listed above. Although only a small 3.9% improvement was obtained for this statistical validation, the contextual classification result looks visually much closer to the ground truth classification, having a much 'cleaner' appearance. We are pursuing more extensive and varied test regions where extensive ground truth data are available. On a larger data set (where only limited ground truth data were available), the time required to perform a 512×390 contextual classification on the MPP was approximately 1 h. This compares favourably with doing the per-pixel MLC on a VAX. The MPP therefore affords us an opportunity to explore spatial classification techniques that until now have been considered too time consuming.

6. Future projects

As pointed out in the introduction, space-borne imaging sensors for remotely observing the Earth will continue to evolve in capabilities and continue to add more massive volumes of valuable scientific data. Already, the planned capability of instruments for acquiring data on the earth-observing system platform of the space station greatly outstrips our ability even to transmit all this data back to our ground processing systems. In addition, it is expected that scientists in the manned space-laboratory modules will want to carry out research campaigns on events of opportunity requiring on-board image processors of extensive performance capabilities. As a result, studies are underway at NASA to explore new technologies required for greater on-board processing capabilities, to capture greater amounts of potentially useful data.

The CMOS (complementary metal–oxide semiconductor) technology that was used in the development of the MPP, for which I have here attempted to demonstrate a unique potential for remote sensing applications, was based upon the latest custom chip capabilities of the late 1970s and early 1980s. The MPP array uses 88 boards each containing 24 chips. On each chip are embedded 8000 transistors allowing the design of 8 complete processors that use circuits of line width 4.5 μm. In addition, adjacent to each chip on the board are two memory chips of 4 K bits each with appropriate pin connections to the processor chips thus providing 1 K bits of memory per processor. Parallel computers with arrays of 16000 to 64000 processors are now commercially available at prices only slightly greater than mini-processors of the VAX-11/780 class. Parallel processors with somewhat smaller arrays and deeper memories are also becoming available at costs comparable with SUN or micro-VAX workstations. Clearly, parallel processing is the likely way to go in the pursuit of new image processing capabilities.

In this regard, NASA supported research studies have already shown that one can produce chips with approximately one million transistors allowing for the design of 128 processors on a single chip with each processor containing over 1 K bits of on-chip memory. Memory chips are also available with 4 megabits allowing for an additional 16 K to 64 K bits of off-chip memory to each processor. Thus, well before 1995 when the space station is expected to be operational, MPP-like workstations in a single board should be available with more power than today's MPP. It is similarly expected that a fully space qualified MPP-like workstation will also be available for research aboard the manned laboratory modules of the space station. In all probability, one could also expect to find at the large ground-processing facilities, parallel processors with arrays of more than a million processors and internal memories of more than a gigabyte.

Thus, the immediate challenge facing the remote-sensing community is not in the hardware arena, but rather the development of greatly improved parallel image algorithms that will exploit these new computing architectures to maximize the extraction and use of the full scientific information content of image data.

I express my deepest appreciation particularly to Dr Strong, Dr Ramapriyan and Dr Tilton not only for allowing me to present elements of their current research at this discussion, but for taking considerable time from their busy work schedules to explain and often tutor me in the methodologies they used.

References

Batcher, K. E. 1980 Design of a massively parallel processor. *IEEE Trans. Computers* **C-29**, 836–840.

Dorband, J. E. 1987 Ray tracing on the MPP. In *Proc. 1st Symposium on the Frontiers of Massively Parallel Scientific Computation, Greenbelt, 24–25 September 1986*. Conference Publication 2478, NASA/Goddard Space Flight Center.

Fischer, J. 1987 In *Proc. 1st Symposium on the Frontiers of Massively Parallel Scientific Computation, Greenbelt, 24–25 September 1986*. NASA Conference Publication 2478, NASA/Goddard Space Flight Center.

Marr, D. & Poggio, T. 1979 A computational theory of human stereo vision. *Proc. R. Soc. Lond.* B **204**, 301–328.

Potter, J. L. 1985 *The massively parallel processor*. Cambridge, Massachusetts: MIT Press.

Quam, L. 1984 Hierarchical warp stereo. In *Proc. Image Understanding Workshop, New Orleans, October 1984*, pp. 149–155. DARPA.

Strong, J. P. & Ramapriyan, H. P. 1987 Massively parallel correlation techniques to determine local differences in pairs of images. In *Proc. 2nd Int. Supercomputer Conference. Santa Clara, 5–8 May 1987*. St Petersburg, Florida: International Supercomputing Institute, Inc.

Tilton, J. C. 1986 Contextual classification on the massively parallel processor. In *Proc. 1st Symposium on the Frontiers of Massively Parallel Scientific Computation, Greenbelt, 24–25 September 1986*. NASA Conference Publication, 2478, NASA/Goddard Space Flight Center.

Discussion

D. Lane (*Intelligent Automation Laboratory, Department of Electrical and Electronic Engineering, Heriot-Watt University, Edinburgh, U.K.*). The hardware architecture of the MPP provides a processing tool that is potentially very powerful. However, its usefulness will depend upon the ease with which users can express the solution to their problem in a machine format. How, therefore, is software written for the MPP, and to what extent does it allow maximum exploitation of the MPP's capabilities?

M. Halem. There is, of course, an assembly language, but in addition there are several fully developed and supported high-level languages with parallel extensions. The most popular and frequently used language is MPP PASCAL in addition to FORTH. An implementation of pure LISP has been demonstrated. The MPP PASCAL is compiled on the host DEC 8800 or VAX 11/780 and FORTH is interpreted while running on the MPP. The languages are also supported with highly efficient built in standard mathematical library routines such as sin, cos, max and min as well as data reordering routines based on sorting. A transportable simulator is also available for use by investigators on their own local VAX/VMS computers for simulating the MPP codes at the hardware level. Development of a parallel extension of 'C' is in progress. MPP PASCAL has been found to be comparably efficient on the MPP to serial compilers on standard architectures.

J.-P. A. L. Muller (*Department of Photogrammetry and Surveying, University College London, U.K.*). Your examples all have one thing in common: the fact that the problem is content- and context-independent so that a SIMD architecture, such as the MPP, is excellent for processing the same algorithm on every pixel in the image scene. It would be useful to know whether mapping less well-behaved algorithms on the MPP has been looked at, and how relevant Dr Halem's examples are to future automated image understanding systems.

M. Halem. A number of less well-behaved algorithms have been mapped and efficiently implemented on the SIMD architecture of the MPP. Fast Fourier transforms are implemented by assigning block sub-arrays of processors to perform FFTS on lines of an image as described in our SAR processing. In the graphic three-dimensional rendering of the scene in elevation mapping, sort computation which are global in terms of routing data to arbitrary processor were required and effected with a two-dimensional bitonic sort. In general, numerous scientists are mapping non-mesh oriented algorithms onto the MPP and other SIMD machines.

Phil. Trans R. Soc. Lond. A **324**, 373–380 (1988)
Printed in Great Britain

The design of environmental geographic information systems

By M. J. Jackson, W. J. James and A. Stevens
Laser-Scan Laboratories Ltd, Cambridge Science Park, Milton Road, Cambridge CB4 4FY, U.K.

[Pullout 1]

The development of geographic information systems (GIS) is recognized as a prerequisite for the effective exploitation of remotely sensed data. Current commercial systems represent a solution with a strong bias from either the mapping *or* the remote-sensing market. They thus lack full cross discipline functionality and a model-oriented approach. This paper examines some of the key issues in truly integrated GIS design.

The five data types of image, object (vector), terrain, tabular and knowledge are identified along with the operations required with them within a GIS. The term 'geoschema' is introduced (analogous to schema within a database) to describe the organization of the geographical datasets. Three-dimensional data handling, the necessity of qualifying data and the user interface are given particular attention. An efficient method of implementing an integrated spatial index into the data sets is described.

1. Introduction

The development of geographic information systems (GIS) is not new; some systems such as the Canadian GIS (CGIS) (Crain & Macdonald 1984) date back over 20 years. There has, however, been a remarkable recent surge of academic, political and commercial interest (table 1). Many of these initiatives have been stimulated by remote-sensing developments. In particular the following factors are relevant:

(i) large quantities of satellite remotely sensed data are now available with good radiometric, temporal and geometric resolution; especially from the *Landsat* thematic mapper and French *SPOT* sensors;

(ii) advances in computer processing and display technology now make practical the manipulation of large quantities of geographically referenced data for commercial applications;

(iii) following the large investments in space programmes, there is growing worldwide political pressure to generate commercial returns to contribute to 'ongoing' programmes.

The development of GIS, and especially those integrating remotely sensed and other data types (so-called integrated GIS or IGIS), is now recognized as a prerequisite for the effective exploitation of remotely sensed data. However, definition of current and future user needs and the consequent software engineering issues underlying an effective design for an integrated GIS have still received only scant attention. This paper, therefore, examines some of the key issues in IGIS design, concentrating on those aspects previously given inadequate consideration.

Table 1. Some recent U.K. initiatives in gis[1]

event	date
House of Lords enquiry into remote sensing and digitial mapping	1983–1984
Department of Trade and Industry funded study on gis	1984–1985
Establishment of European Association of Remote Sensing Laboratories Working Group on Integrated gis	1984
Advisory Board for Research Councils identifies spatial data handling techniques as justifying research priority	1984
Natural Environment Research Council special initiatives in gis (includes 3 university contracts)	1984
European Space Agency sponsors study of igis	1985
EEC commences the corine gis programme	1985
U.K. National Remote Sensing Centre gis Group established	1985
Committee of Enquiry on geographic data handling	1985–1987
British National Space Centre awards gis contracts	1987

[1] Also European initiatives with U.K. involvement.

1.1. *Definition of gis and igis*

A brief working definition of a gis for the purposes of this paper may be 'a computer based system for the efficient input, storage, manipulation, analysis, representation and retrieval of all forms of spatially indexed and related descriptive data'. More complete definitions may also be found (Wilkinson & Fisher 1986; Smith *et al.* 1987).

Although this definition can embrace systems incorporating multiple types and formats of data, e.g. point, line and polygon vector representations, raster data and tabular statistics, there has historically been a separation between vector, raster and non-geographical statistical systems. It has therefore been useful to introduce the igis acronym to explicitly refer to a gis integrating the three data types.

1.2. *Current systems*

The simplest, and still the most common configuration for a gis that incorporates remotely sensed and map inputs, is two discrete systems implemented either on separate or common computer systems. Data are passed between the two systems by vector–raster and raster–vector software. However, despite this software link, the systems often remain as separate cartographic gis and image analysis systems, with only minimum integrated software development and use of the different datasets. Any operation combining raster and vector data is forced into one domain to the detriment of the other datatype. There are a number of effects: loss of attribute, topological and context information; loss of spatial resolution; introduction of data redundancy; limited interaction between the vector and image data; maintenance of both a vector and a raster database; two distinct user interfaces are maintained.

Where the two systems are on distinct hardware, there are additional disadvantages: the physical separation of the systems results in extra time needed to combine datasets; system upgrade is complicated by the parallel and often independent development of two discrete systems; the need to maintain two independent systems incurs a considerable cost penalty at time of purchase, in user training and in maintenance and support costs.

The single workstation system represents the current commercial state of the art. All represent a solution with a strong bias from either the mapping market *or* from the remote sensing market. They thus lack full cross-discipline functionality and a model-oriented approach (as distinct from data oriented).

1.3. *Hybrid solutions*

A better solution to the GIS requirement is to provide simultaneous access to independent raster and vector datastructures and to handle the integration at a higher level. This approach is taken by Laser-Scan in its terrain validation and exploitation system (TVES) which allows the user to open and interact with separate raster and vector data files displaying registered results on a single display. The user can then manipulate map data as features (roads, rivers, woods, etc.) and terrain or images as raster data, while benefiting from an integrated view of the whole.

2. Technical requirement for future IGIS

Current IGIS design is based on *ad hoc* mechanisms of data exchange between systems or on a more integrated approach but tailored for a specific application area. In addressing the design issues for a general purpose IGIS, the following technical requirements have been identified.

2.1. *IGIS data types*

Geographic data exist in a wide variety of forms, but these may be grouped according to their *type*. Each data type may require distinct structures and methods of handling within the IGIS. The data types are: image data in raster form, for example, satellite imagery; object data in vector form, for example, digital cartographic data; terrain height data, commonly digital terrain models, contour–spot-height maps or surface models; tabular data, especially attributes, indexes, or references; rule–knowledge data, for support of decision making processes.

2.2. *IGIS operations and features*

The following list is *not* exhaustive, but includes those operations most vital to an effective IGIS.

Data input: format conversion programs, to take 'popular' digital datasets, and published interfaces, allowing users to create their own input programs; digitizing mechanisms, providing a route to input documents, and text input facilities.

Data edit: efficient, data type specific editors, for use during input and update of datasets; multidata-type editors, allowing editing in the context of several data types.

Data storage: storage of data in their 'natural' structure, avoiding data degradation; qualification of data, description of the content, characteristics and origin of any dataset; access control to datasets through ownership, and through partitioning; relation of datasets via geoschema (see §3.1).

Data access: data catalogue, which itemizes all available local and remote datasets; search and select facility, which locates datasets according to user parameters.

Data preprocessing: image processing facilities; image geolocation and classification, by reference to object and knowledge data; object geolocation and transformation, by reference to tabular or object data; object coding and definition, perhaps by reference to image data.

Data output: format conversion programs to popular datasets, and published interfaces, which allow users to generate their own output datasets; availability of qualifying data for output datasets.

Data display and plotting: display and plotting of combined data (e.g. 3D views comprising image, object and terrain data); generation of cartographic products; availability of display hardware optimized for image and object display.

Utilities: the primitive operations required by the user for extraction include: spatial association of datasets, which is based on their geoschemas and automatically applies transformations to convert between geoschemas; combination of datasets, including spatial overlay and intersection, to generate new datasets if required; search and selection of regions and objects on spatial and attribute criteria; calculations, report generation, display and plotting facilities; conversion between data types.

User interface: menu or choice driven interface, that advertises and gives access to all available facilities; data catalogue browse, which has search and select mechanism that finds relevant datasets; project definition, which specifies geoschema(s), storage partitions, procedures, etc.; geoschema definition, which specifies spatial reference system, indexes, attributes, etc.; procedure definition, which specifies sequences of primitive operations; problem definition, which generates procedures that reduce the need to understand the operation of the system.

System configuration: there are a number of desirable requirements of the IGIS design, that impinge directly on its implementation in a given processing environment: the IGIS user interface and data-handling concepts should be invariant from system to system; the IGIS software structure should be modular, allowing subsets of functionality to be implemented according to the facilities offered by the computing system; the IGIS functions themselves should have variants to take advantage of local computing facilities.

3. IGIS DATA ORGANIZATION

In meeting the technical requirements identified above, the major challenge lies in designing the structures for the data organization. These must permit the maintenance of data in its primitive input form to preserve precision and information content, while offering the user an integrated view of the whole data environment without involving data transformation or registration. In addition, the structures must provide fast and efficient spatial processing. A schematic diagram of the organization proposed is shown in figure 1. The details of this data organization are described in Jackson *et al.* (1987). The four following specific topics, however, justify comment.

3.1. *Geoschema*

This term is analogous to the use of 'schema' within a traditional database. It describes the organization of the geographical datasets. In particular it defines the spatial referencing system, indexes, classification and attribute coding schemes. The enforced presence of such information for each dataset permits the IGIS to automatically relate them by transforming, resampling or translating as required. Benefits in efficiency accrue from relieving the user of this responsibility,

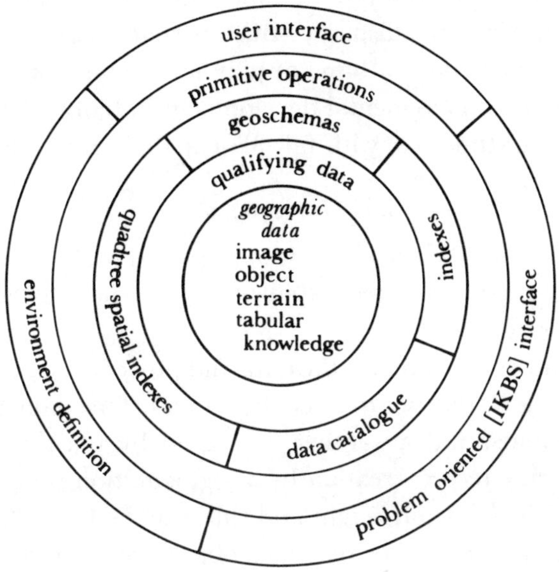

FIGURE 1. Diagram of integrated GIS.

although further gains may be made if all the data are deliberately coerced into the same geoschema. The user can concentrate more on the content of the datasets, rather than their representation.

3.2. *Data handling*

Absolute height, slope, aspect, range and terrain roughness are necessary parameters in considering many environmental issues. Examples include the distribution of selected flora and fauna, modelling snow-melt or run-off characteristics, geological interpretation and cross-country mobility studies. The surface form is also important to image classification, particularly with the greater look angles which are obtained with TM, off-nadir *SPOT* data, SAR images, airborne scanner data, etc. Until now, remote sensing and cartographic GIS systems have largely been two dimensional because of the cost and technical limitations. As a result, vector, raster and quadtree based systems have neglected the third dimension in both practical implementation and theoretical development.

The solid modelling and 3D systems used in CAD would not appear to be very applicable to 3D GIS (Besl & Jain 1985). Many of these are based on constructive solid geometry where an object is built of primitive shapes such as cones and cylinders. This is inappropriate for representing terrain. Another approach, and one favoured by military simulator systems, is the use of surface boundary representation. This is a more favourable approach for terrain and includes triangular as well as polygonal approaches.

Despite the increased complexity of 3D systems, a future trend towards representation and display of the third dimension within IGIS systems is anticipated. Data are becoming more generally available as a result of military programmes, and the increasing power of workstations now makes the computations practical. In addition, many new hardware developments for 2.5D and 3D vision are planned. These include an Alvey project (Alvey MMI137 (Muller, this symposium, p. 381)) involving the transputer (Inmos 1986) as a flexible general-purpose vision engine, and more specific graphic displays (BCS Systems 1987). Such devices offer the computational power necessary for real-time display and animation.

From a design viewpoint, an incremental development initially offering 2.5D surface representation would seem most attractive, with full 3D capabilities to follow.

3.3. *Qualifying data*

The importance of qualifying data has also been neglected in the past, particularly in image-based GIS. These data accompany any geographical dataset and may generally be divided into historical and status information.

The historical qualifying data references the data and procedures from which the dataset was created and modified. It allows the user to trace the sources of the data. Each record will specify a procedure, the data inputs, and any other data used by the procedure. For example, a classified satellite image may record creation by a segmentation process that used a specified source image, reference object, terrain height and rule data, followed by modification through a classification process that used further reference object data and a different set of rules.

The status information comprises data in addition to the content of the dataset (often termed 'header data' because of their location in the file). It includes geographical coverage, description of content, scale, accuracy, currency, etc. Some of this will be required by processes that manipulate or interpret the data.

Qualifying data is essential when bringing together geographic data from different sources, or when distributing data to a wide user base that had no involvement with their generation. It will be particularly relevant as greater use is made of knowledge or GIS-assisted techniques of image classification. It should be the primary input to error-train analysis, or to expert systems 'advising' on what is appropriate when combining datasets.

3.4. *Spatial indexing*

Much of the research effort to date on integrated GIS has concentrated on low-level data structures (such as the quadtree (Samet 1984)), which can hold both vector and raster data in a uniform way. Although this use of quadtrees may be justified for certain types of operation, especially in the context of predefined projects, they do not offer a sensible general purpose solution to IGIS integration. However, they do have benefits in providing an integrated spatial index and as such play a major role in the design proposals presented in this paper.

At the highest level of area storage, the operator is aware of one common spatial index into the data. At 'lower' levels the index contains references from the structure into the particular (raster or vector) datasets. Obviously, the design and implementation of the spatial index and the underlying datastructures have a significant effect on the speed and flexibility of the IGIS system.

Although a detailed description of integrated spatial addressing schemes is beyond the scope of this paper, the approach identified may be summarized as follows.

Vector data are stored in a geometry file and spatially indexed by a PM3 quadtree (Samet *et al.* 1985). Quadtree nodes are successively subdivided until there is a single data point at each leaf. Any lines passing through the leaf's area (connecting other data points) are also recorded with the leaf data. An example of a portion of a vector map decomposed in this way is presented in figure 2. The quadtree is implemented as a pyramid and intermediate levels of

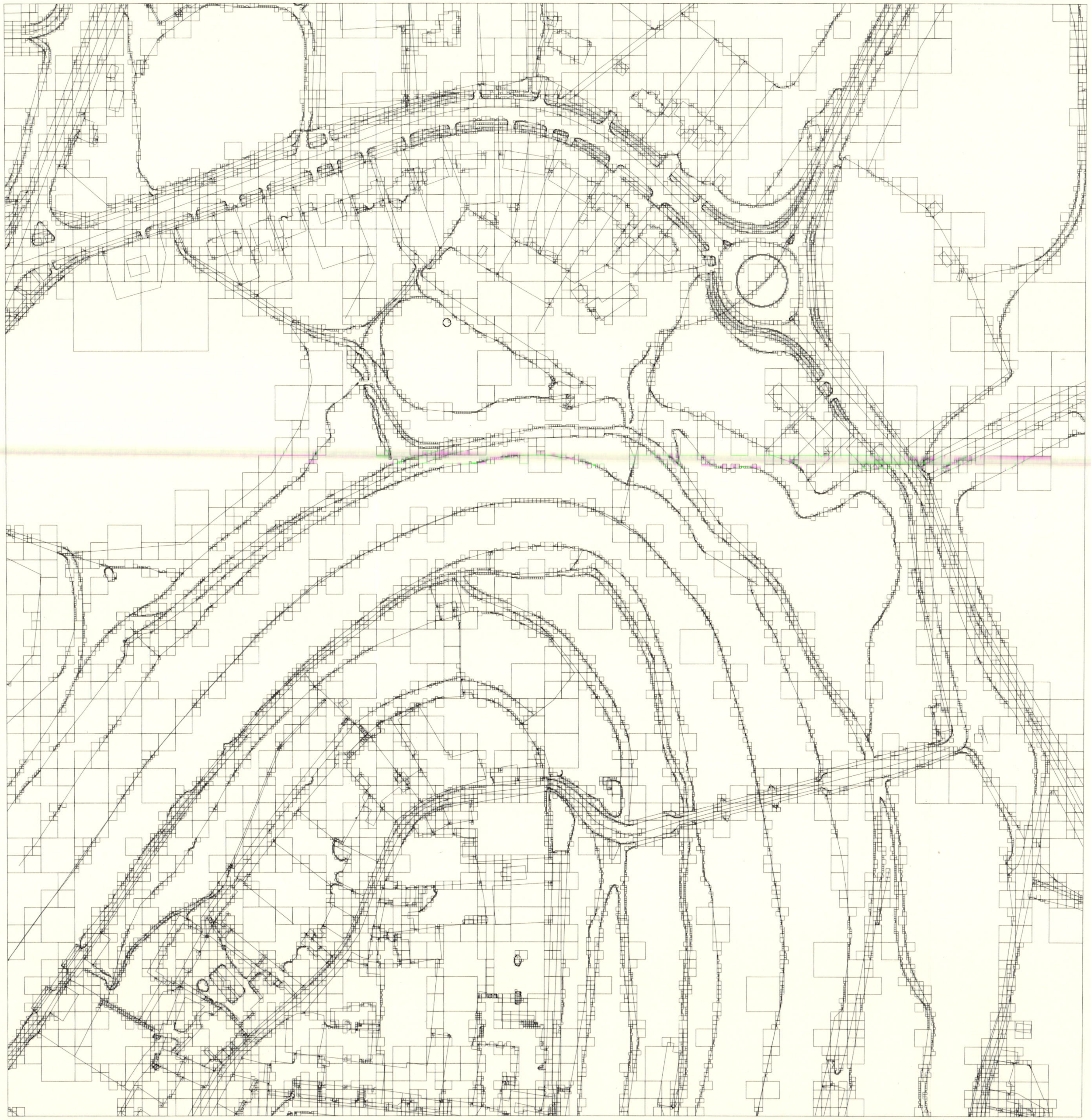

FIGURE 2. Example of PM3 decomposition of a map.

(*Facing p.* 378)

the spatial hierarchy contain summary information about features referenced by the lower levels. The quadtree is truncated at a predetermined level for storage efficiency but all the vector information is still represented at its original spatial resolution.

Raster data of low-dimensional order (such as binary data) may be stored in a region quadtree. However, most raster data, and especially those derived from remotely sensed images, are highly dimensional (e.g. several bands of 8-bit information) and cannot be efficiently held in a quadtree. For these data, a better option is a form of grid file with data stored in a regular patchwork of local raster 'tiles'. A hierarchical spatial index in the form of a quadpyramid can be superimposed on the tiled data to provide summary information at its higher levels. For pixels that have been registered to vector information, the two quadpyramids can be chosen such that the areas covered are identical and that the vector quadtree has at some level a node representing the same area as the raster pixel.

4. Conclusions

Much of the abundant literature on GIS, has concentrated either on relatively simple discrete case studies, on generalized discussion of GIS applications or on specific aspects of the low-level data structures for integration. For effective system design, the detailed structure and organization of the entire system is critical, and the whole IGIS environment must be considered, from the moment a magnetic tape is loaded onto the system, through to final plot, display or report. In particular, the use of qualifying data is important to understand the pedigree and reliability of any output, and the IGIS metadata is vital to the organization of the IGIS system.

As functional capabilities increase, making use of 3D data sets, IKBS techniques and special purpose hardware, the need for an integrated approach to GIS design will become mandatory to control the large volumes of data involved. The approach presented here is a flexible and upgradable route towards such a system.

References

BCS Displays 1987 *3D Graphics hardware*. BCS display group documentation 1987. British Computer Society London.

Besl, P. J. & Jain, R. C. 1985 Three-dimensional object recognition. *ACM Comput. Surv.* **17** (1), 75–145.

Crain, I. K. & MacDonald, C. L. 1984 From land inventory to land management: the evolution of an operational GIS. *Cartographica* **21**, 40–46.

Inmos 1986 The transputer family. Product Information.

Jackson, M. J., James, W. J. & Stevens, A. 1987 British National Space Centre GIS demonstrator feasibility study, BNSC Report 1987.

Samet, H. 1984 The quadtree and related hierarchical data structures. *ACM Comput. Surv.* **16**, 1–67.

Samet, H., Rosenfield, A., Shaffer, C. A., Nelson, R. C., Huang, Y. & Fujimura, K. 1985 *Applications for hierarchical datastructures to geographical information systems*, Phase IV. Computer Science Technical Report, University of Maryland.

Smith, T. R., Menon, S., Star, J. & Estes, J. E. 1987 Requirements and principles for the implementation and construction of large-scale geographic information systems. *Int. J. geogr. Inform. Sys.* **1** (1), 13–31.

Wilkinson, G. G. & Fisher, P. F. 1986 *Review of integrated geo-information systems techniques*. Working Group 13 of the European Association of Remote Sensing Laboratories.

Discussion

C. W. MITCHELL (*Department of Geography, University of Reading, U.K.*). Geographical Information Systems manipulate overlaid data sets of a number of different themes such as topography, geology and land cover, the data being referenced by grid coordinates rather than by recognizable land features. The effect of this is that there is no clear mechanism for extrapolating thematic data between environmentally analogous, but geographically separated, areas. An established method of achieving this is to recognize analogies between the recurrent natural units of the terrain (which integrate soils, water, vegetation, etc.) on remotely sensed imagery. The ability to make and test such extrapolations is an essential part of the environmental sciences.

How is it possible for a GIS of the sort described by Dr Jackson to provide this extrapolation capability?

M. J. JACKSON. The requirement to match spatially separate areas in terms of their geographical characteristics for prediction has traditionally involved the input of a high level of interpretative skills. The IGIS by providing a computational environment whereby some of the interpretation and extrapolation may be more easily quantified should greatly assist the analysis in terms of accuracy and consistency of prediction. Geostatistical techniques based on spatial correlation, principle component and factor analysis will also be aided. Intelligent knowledge-based systems (IKBS) or expert systems technology may, in addition, allow some of the empirical processes of the analyst to be automated, though the degree of interpretative skills now used by the specialist is unlikely to be matched in the short term, especially as they are based on the poorly understood abilities of the human to generate 'like areas' out of complex spatially interacting data.

D. LANE (*Intelligent Automation Laboratory, Department of Electrical and Electronic Engineering, Heriot-Watt University, Edinburgh, U.K.*). Techniques for representing three-dimensional shape within a computer have been studied for some time by researchers in the computer graphics and computer vision communities. It has resulted in techniques using polygons, parametric surfaces, octrees, the $2\frac{1}{2}$D sketch and so on. How are researchers in the remote sensing community modelling three-dimensional terrain within Integrated Geographic Information Systems?

M. J. JACKSON. Many of the 3D shape representations such as constructive solid geometry (CSG) and octrees have been successfully used in CAD-type systems. For terrain visualization the CSG approach is not efficient because of the generally gently curving nature of the surface. Octrees, by analogy to quadtrees, could be used but are really a 3D volume rather than $2\frac{1}{2}$D surface representation. Although efficient for volumetric computations they will not be particularly good for rotation, translation and scaling.

Current terrain methods are generally based on the Delaunay triangulation. Height values are assigned to each vertex and the triangular surface facets represent the slope and aspect of the local regions.

With advances in hardware to manipulate polygonal facets rapidly in 3D, future visualization methods may explore this approach. Computing advances also open up possibilities for ray tracing and fractal generation to improve surface realism.

Phil. Trans. R. Soc. Lond. A **324**, 381–395 (1988)
Printed in Great Britain

Key issues in image understanding in remote sensing

By J.-P. A. L. Muller

Department of Photogrammetry and Surveying, University College London, Gower Street, London WC1E 6BT, U.K.

Remotely sensed images of a planet's atmosphere, oceans and surface contain a plethora of confusing signals about the physical nature of these phase states. Historically, there has been an emphasis on the semi-automated extraction of feature classes based on the spectral properties of objects viewed within a scene and on the use of *ad hoc* manual photointerpretation techniques. Although these approaches will remain important, they are inadequate, on grounds of speed, accuracy and cost, for the increasing demands of data-gatherers and consumers.

Research has recently begun into the automation of image-interpretation tasks and the development of parallel machines with the required processing capabilities. Three important requirements are: (i) means to simulate the appearance of a scene, including the interaction of electromagnetic radiation with the surface and the effects of any intervening atmosphere; (ii) an understanding of how knowledge can be captured and introduced at different levels in the processing hierarchy and (iii) the application of constraints based on a knowledge of the geometry of objects in the scene. These three aspects will be illustrated by examples from various fields, including petroleum exploration, measurement of fluid motion and the extraction of digital terrain elevation models.

1. Introduction

Image understanding is concerned with the recovery of physical properties of a scene from image features and the computational systems that interpret images (see, for instance, Levine 1978; Tenenbaum *et al.* 1980; Marr 1982; Herman & Kanade 1986; Tailor *et al.* 1986). Remote sensing involves the extraction of physical properties of fluids or solids and their interpretation for a specific application end use, such as cartography (Doyle 1982) or climatology (Bretherton 1985). Image understanding in remote sensing is defined here as the development of techniques and computational systems for the automated extraction of scene properties from satellite and aerial imagery for specialist domains.

Requirements for such an automated extraction system arise from the increasing demands of data consumers (see, for instance, Goetz *et al.* 1983; Conway & Browning, this symposium) for more objective and quantitative information promised by a new generation of optoelectronic and microwave digital data gatherers (see, for instance, Goetz *et al.* 1985). Remotely sensed imagery is also believed to have a potential major role in spatial information systems under current development for geographical and environmental applications (see, for instance, Jackson, this symposium).

Automation has long played a role in the processing of satellite imagery (see, for instance, Jones & Colombeski 1981). However, the tendency has been to restrict automation to either preprocessing tasks, requiring a minimum of knowledge external to the image (e.g. associated telemetry-based estimates of the satellite position), or to crude attempts to force correspondence between image patterns (e.g. cloud-tracking with pairs of geolocated geostationary satellite images: see, for instance, Morgan 1979). The degree of automation has frequently been a function of the perceived needs for volatile operational products and the efficiency of machine processing compared with human manual intervention. Examples include the production of cloud images for weather forecasting and, in the field of Earth resources, the use of *Landsat* data in the LACIE system for wheat-harvest monitoring and forecasting (NASA 1979).

The LACIE system is a prime example for a new trend in remote sensing, that of trying to create a combined information processing system, based upon a desire to understand physical processes. It included some degree of automation in the production of wheat-field coverage from classified images (by using manually selected training sets in the original multispectral images) and the application of knowledge external to the image (for instance, contemporary weather data) to predict harvest yield. This system was criticized for the lack of site-specific information (Swain 1985) but was the first major application of remote sensing applied to earth resources that used computer analysis for very large areas.

2. Limitations of current techniques

Current practice in geographical applications of remote sensing still relies heavily on the use and further development of statistical procedures (e.g. supervised or unsupervised classification of multispectral data) to try to extract information on surface cover based upon an attempt to separate ground-cover types by their spectral signature, even though there have been strong arguments advanced against its continued use (see, for example, Tenenbaum *et al.* 1980; Gerstl & Simmer 1986).

It is now widely recognized that multispectral classification has a number of severe shortcomings (Swain 1985); these include the following.

Signature extension. There is difficulty in extrapolating local spectral characteristics to larger geographic areas; this is both a sampling problem and a fundamental limitation in the use of laboratory spectral data in the field.

Mixture pixels. Every satellite-image pixel is likely to have more than one ground-cover type within the pixel's field of view.

Signature confusion. The atmosphere and the surface slope (see §3.3.) contribute part of the radiance recorded in every pixel.

Signature ambiguity. Owing to technological constraints (see Goetz *et al.* 1985) spectral bandpasses of sensor channels are frequently too broad to differentiate between ground-cover types (see Bernstein (1986) for a justification of *Landsat* Thematic Mapper band selection and Gerstl & Simmer (1986) for a critique of the severe limitations of relying on spectral signature for vegetation type discrimination).

Temporal instability. Multitemporal classified satellite imagery is both a function of signature confusion (namely the atmospheric conditions change over time) and the observation that spectral signatures for many ground cover types (namely vegetation) change over the time period between satellite overflights.

The difficulties of correcting for geometrical instabilities of aircraft have severely restricted the operational application of digital sensor technology to environmental monitoring. However, the use of aerial photographic data, begun in 1859 with Gaspard Tournachon's balloon-based observations, has continued unabated until now. Photogrammetric techniques (see, for instance, Slama 1980) have been developed to correct for geometrical distortions associated with aerial photography, permitting manual stereo measurements of plan and elevation to be made with accuracies greater than 1 part in 10000 of flying height (Slama 1980). However, although computer assistance is now given in analytical plotters to permit online computation of geometry, little if any automation has been introduced into either the manual stereo measurement or the manual photointerpretation process. A notable exception for the latter is the combined use of map-derived data with digitized photographs (given a good camera model) in automated aerial image interpretation (see McKeown *et al.* 1985). However, this process still relies on the manual determination of the camera geometry.

Given this context, this paper attempts to present a radically different approach based on recent advances in machine vision. We address the question of what the key components of an image-understanding system are and report on recent results to develop these components using advances in computer graphical techniques, data structures and parallel processing technology.

3. Key components

3.1. *System architecture*

A simplified diagram is shown in figure 1 of the key components of a hypothetical image-understanding system in remote sensing. Missing from this diagram are explicit indications of the possible data flows and the invocation of (implicit) knowledge sources (see later).

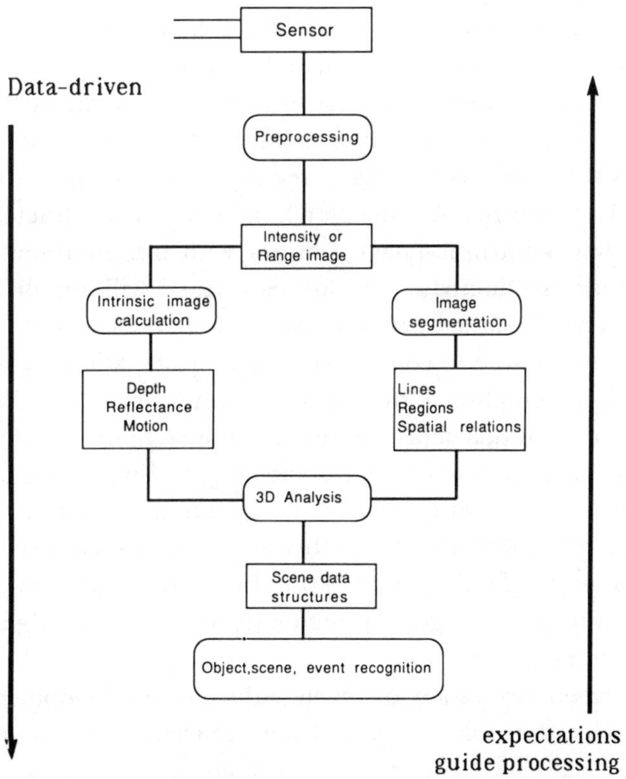

FIGURE 1. Key components of an image-understanding system for remote sensing. Such a system can be developed either for bottom-up control or the use of expectations to guide processing. Crucial features of such systems are the incorporation of three-dimensional information and the attempt to include knowledge about the image formation process.

There are two basic modes of operation of the system, commonly referred to as bottom-up or data-driven (i.e. from the sensed data to a scene description) and top-down or expectations-guide-processing (i.e. search for a specific instantiation of a physical object(s) in the sensed data).

Computer vision research (referred to as 'high-level' vision, see Havens & Mackworth (1983)) is usually concerned with top-down processing (Binford 1982). It can also be applied to aerial imagery to search for specific instances of, for example, houses in a suburban estate (see Nagoa & Matsuyama 1980; Matsuyama 1987) or in satellite imagery to monitor or track specific natural or anthropogenic features pre-extracted from maps, given a good camera model (see Tenenbaum *et al.* 1979).

Pattern recognition or image analysis research (referred to as 'low-level' vision, see Havens & Mackworth 1983) has usually been concerned with bottom-up processing such as line extraction (see Brown *et al.* (1983) for an example of a specialized feature detector known as a Hough transform) or region extraction (see Cross & Mason (1985) for an example of split-and-merge tactics) for later matching with a model pattern.

Whichever data or control flow is implemented, the most significant feature of image-understanding research is the inclusion of three-dimensional information based on the realization that it not only affects aerial photography (Herman & Kanade 1986) but, with increasing spatial resolution for satellite sensors, even orthographically projected *Landsat* images (see Justice *et al.* 1981; Woodham & Lee 1985).

Intrinsic image calculation refers to the extraction of depth, both relative (see, for instance, the 'shape-from' methods of Horn 1977) and absolute; the subsequent extraction of reflectance, an invariant physical quantity (to illumination and viewing conditions, see §3.3) by using depth and slope and, for a temporal sequence, motion estimates (for fluids, see §2) or change detection (see Tenenbaum *et al.* 1979). The resultant intrinsic images can then be segmented with a series of low-level operators to recover edges, regions and their spatial relations.

Three-dimensional analysis refers to the matching of features extracted from a combination of intensity and processed intrinsic-image features with instantiations of features extracted from the object and/or scene database. The former features will usually be referred to surface-based measurements and the latter will tend to be volume-based data. The objective of the scene data structures is therefore to convert volume representations to surface ones before 3D analysis. This may be accomplished by the incorporation of surface boundary information directly in volume representation schemes (see Chien & Aggarwal 1986).

Examples of this 3D analysis include map-image registration for the absolute orientation of *SPOT* data (Chevrel *et al.* 1981) with ground control points automatically extracted from digital map data (see Muller *et al.* 1987) and the quality assessment of digital elevation models (DEMS) derived from overlapping *SPOT* images (see Muller & Day 1987) with either manual photogrammetric measurements of geomorphometric features (e.g. ridges, stream networks) or DEMS derived from interpolated contours.

In the next three sections, examples are given of the recent development of some of these key components; in each, the role of knowledge sources and its multilevel invocation are shown.

3.2. *Knowledge sources and encapsulation*

These sources will include models of the image formation process; geometrical characteristics of objects within the scene and their interrelations (see Havens & Mackworth 1983); reflectance characteristics of objects within a scene (see, for example, the DMA (U.S. Defence Mapping Agency) culture file description in Schachter 1980) and specialist expertise. This last source can be simplistically considered as consisting of two main types (see Reid *et al.* 1985): engineering models (e.g. of how oil is formed) and experiential 'rules of thumb' which can be readily combined into a set of rules for an expert system (e.g. applied to mineral prospecting, see Reid *et al.* 1985).

Knowledge encapsulation is still the most time-consuming and error-prone part of any image-understanding system. For example, digitization of paper maps, although automated for certain stages of the mapping process, is still very time-consuming and expensive (see discussion in Dowman & Muller 1986). Similarly, conversion of specialist knowledge into machine-readable rules is fraught with difficulties. For spatial information, attempts must be made to preserve the original accuracy of the data. Appropriate data structures to enable the integration of these multiscale data-sets are currently needed (see Jackson, this symposium).

An example of knowledge encapsulation studied here is raster to vector conversion, which may be required for the automated interpretation of satellite images of the oceans. Current semi-automated approaches rely upon use of the satellite orbital information and manual intervention to locate three or more ground control features (see Ho & Assem 1986). In the approach adopted here, automated edge detection is used to try to extract the coastline based

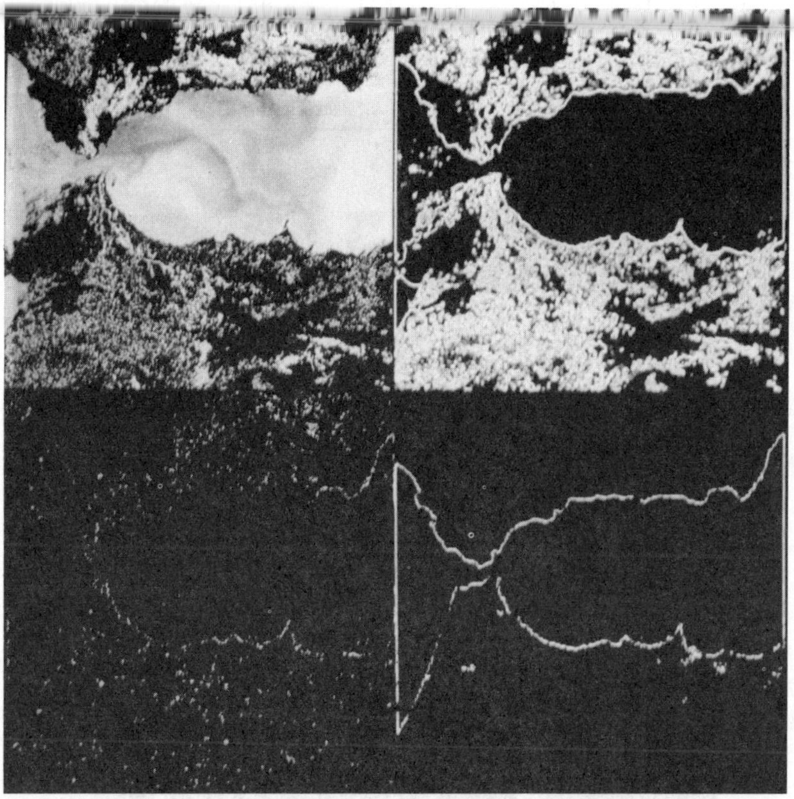

FIGURE 2. Example of automated coastline detection. The upper left figure shows the original daytime NOAA AVHRR image, with the upper right the application of edge detection using a Sobel filter. The lower left shows the application of a texture segmentation based on histogram thresholding of bimodal distributions of standard deviation of intensity within small window areas. The lower right shows the result of this edge-detection after the combination of the other two processes and erosion and dilation.

on gradients and differences in texture between the land, clouds and the sea. The target edge pixels are then connected by an iterative scheme of erosion and dilation (see Preston & Duff 1984). Figure 2 shows the results for a daytime NOAA AVHRR image of the Straits of Gibraltar.

Given a single successful segmentation, it was decided to apply this technique to several successive day and night images. Figure 3 shows the result compared with a digitized map of the area, which shows that the technique is quite robust for daytime images only. Hence, coastline detection must also rely on finding a unique set of correspondences between a

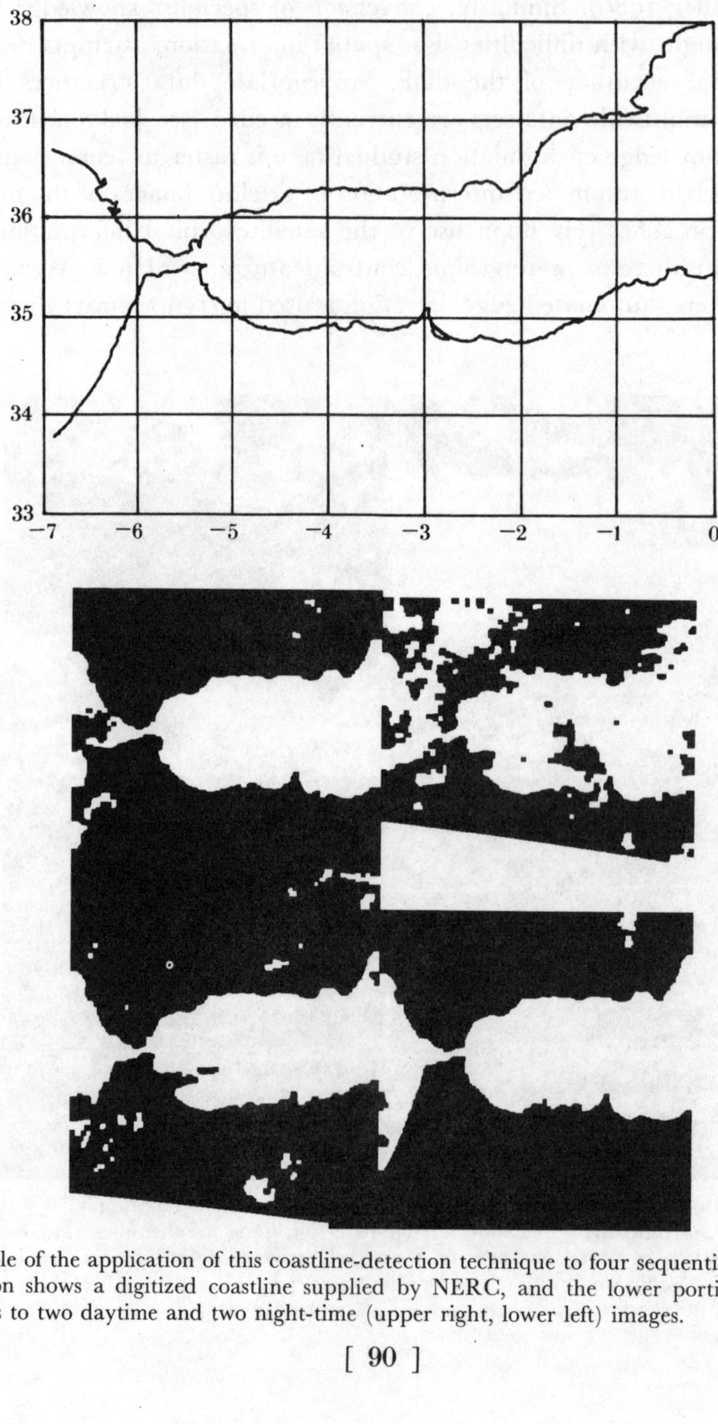

FIGURE 3. Example of the application of this coastline-detection technique to four sequential satellite images. The upper portion shows a digitized coastline supplied by NERC, and the lower portion shows the effect of applying this to two daytime and two night-time (upper right, lower left) images.

'perfect' model and noisy data. Fortunately, recent advances in dynamic programming (see Maitre & Wu 1986) indicate encouraging results in this area.

This example illustrates that several different types of knowledge source and encapsulation may need to be incorporated in automated processing, namely: platform and sensor models; digitized maps; distance metrics as cost function for the dynamic programming; physics (clouds are colder than land); and heuristics (sea is less textured than land).

3.3. *Models for image formation*

Figure 4 indicates schematically the primary sources of radiation received by a passive sensor onboard a satellite, ignoring the strong effects associated with the bidirectional viewing and illumination geometry (see Gerstl & Simmer 1986). It shows that to invert the signal received at the satellite to extract meaningful information on the reflectance properties of the surface cover will to a large extent depend on the relation between geometry and radiometry. Determining whether topography or surface cover dominates the signal received is one of the key issues in this area.

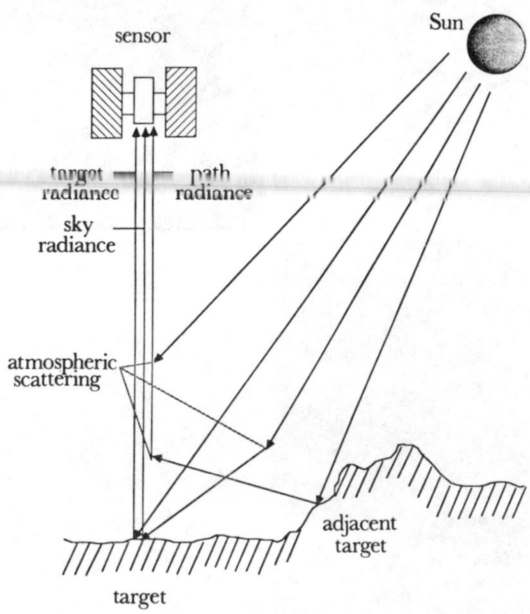

FIGURE 4. Some of the scattering processes involved in passive remote sensing at visible and near-infrared wavelengths (after Woodham & Lee 1985).

The techniques for modelling the effects of all the physical processes involved in scattering incident sunlight into a satellite sensor have recently been tackled with the advent of supercomputer-based simulation techniques. However, the atmospheric component can still be considered as intractable given the limited information available on the local distribution of water vapour and aerosol scatterers (see Pearce 1986 for an attempt to address this question by using mean climate data). Currently, only empirical methods can be used to assess the atmospheric components (path and sky radiance) of the signal (see Teillet 1986) in an attempt to remove their influence.

A simple shading model assuming lambertian reflectance can be applied to model the dominant effects of the surface shape (see Horn & Bachman 1978). Ray-tracing (see Fujimoto *et al.* 1986) can then be used to investigate the effects of multiple scattering and variations in surface reflectance. Figure 5 shows an example of applying a simplified variant of ray tracing to a digital elevation model (DEM) by using just the first ray intersection to model cast and self shadows. The resultant areas which are in cast shadow can then be used in various schemes (see Teillet 1986) to estimate the effects of atmospheric scattering automatically. Surface geometries, such as tree canopies, may also be simulated by either image models (see Luttrell & Oliver, this symposium) or full stochastic models to allow ray scattering to stimulate surface bi-directional reflectance distribution functions (see Cabral *et al.* 1987).

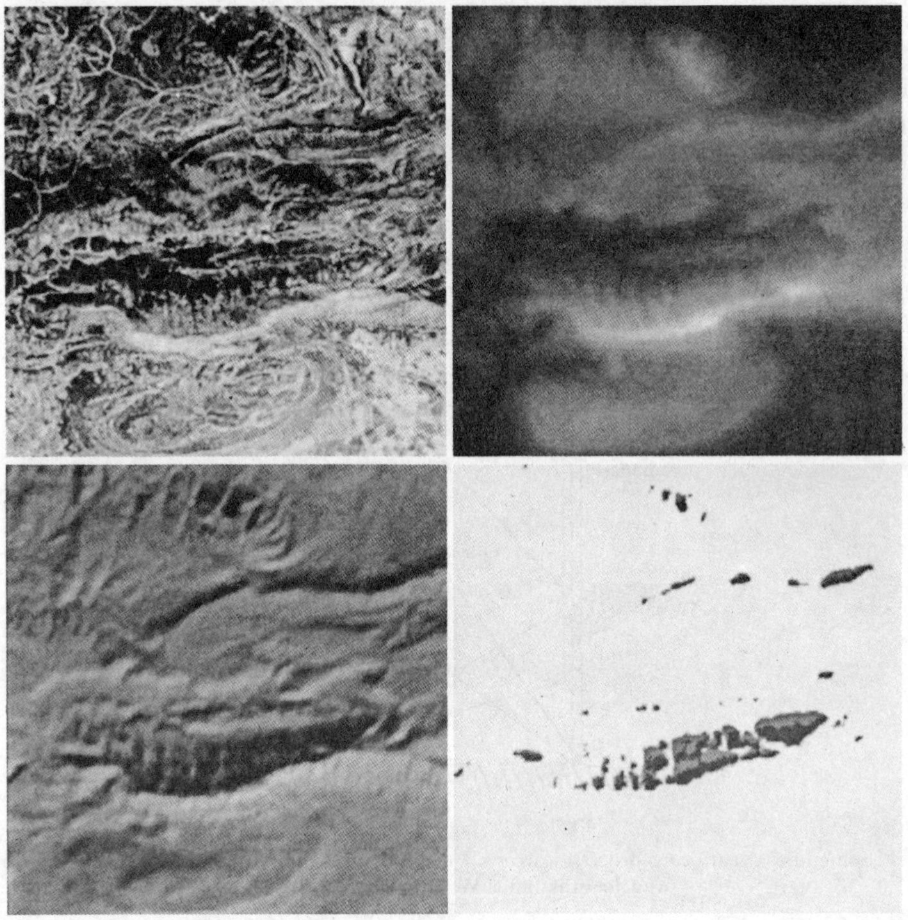

FIGURE 5. The generation of synthetic images for atmospheric correction is illustrated here. The upper left figure shows a small portion of a *SPOT* image (copyright, CNES) of 10 m resolution. The upper right shows an intensity range image of the DEM (corresponding digital elevation model); the lower left shows the simulation of the *SPOT* image with cast shadows included. The lower right shows those portions detected automatically which are in cast shadow that can be used for the correction of atmospheric effects.

These simulations of satellite images can also be used to develop better feature detectors, able to discriminate between effects due to relief and those due to variations in surface cover. Figure 6 shows an example of a comparison of the application of a zero-crossing filter to an original image (processed to 'remove' relief) with the cast shadow model. Developments of this kind may lead to a revolution in the way image processing operators can be developed, by simulating the known physics of the image formation before applying some appropriate heuristics.

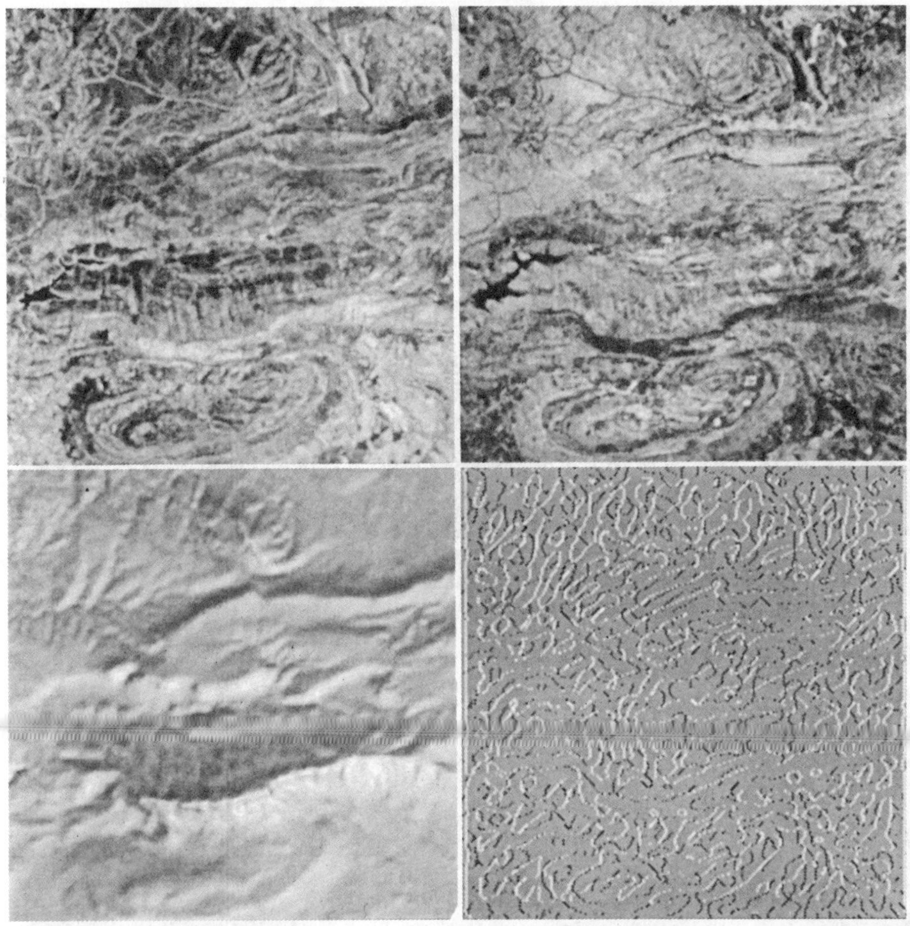

FIGURE 6. Application of synthetic image to the development of machine vision components. The upper left figure shows a *Landsat* band 7 image; the upper right shows a band ratio (band 3 : band 4) which tries to suppress effects of topography. The lower left figure is the same as the lower left in figure 5. Finally, the lower right figure shows the edges extracted from the synthetic image (in white) and those from the band ratio (in black).

This example again shows that several different types of knowledge need to be incorporated in automated processing, namely: platform and sensor models; digitized map data (from DEM creation); physics of image formation.

This technique is currently being tested for its routine application to satellite image processing when trying to detect oil and gas seepage-related events in surface reflectance (see Rock 1984) and for aiding geologists in their understanding of landscape processes in difficult terrain (see figure 7).

3.4. *Knowledge invocation*

Classical image understanding systems generally only try to introduce knowledge at high levels (e.g. object models, spatial relations between objects, see Besl & Jain (1985)) although there have been some recent attempts to introduce knowledge at low levels also (see Nazif & Levine 1984).

Little work, however, appears to have been done on trying to introduce knowledge at different levels explicitly. We will take two examples of our work: automated fluid motion measurement and automated stereo matching.

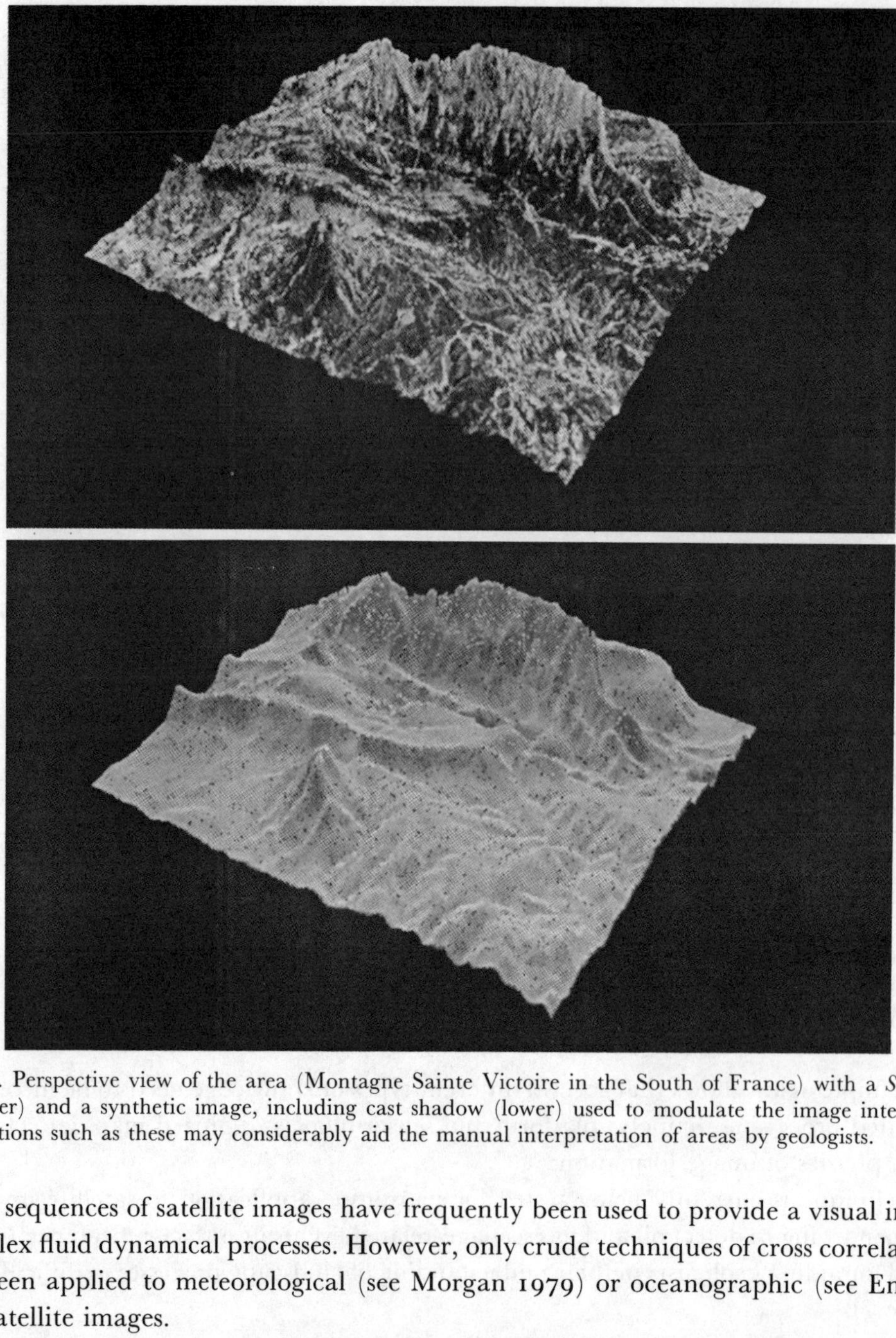

FIGURE 7. Perspective view of the area (Montagne Sainte Victoire in the South of France) with a *SPOT* image (upper) and a synthetic image, including cast shadow (lower) used to modulate the image intensity. Visualizations such as these may considerably aid the manual interpretation of areas by geologists.

Time sequences of satellite images have frequently been used to provide a visual impression of complex fluid dynamical processes. However, only crude techniques of cross correlation have so far been applied to meteorological (see Morgan 1979) or oceanographic (see Emery *et al.* 1986) satellite images.

Optical flow (see Horn & Schunck 1981) can be used for motion detection, although its original application was rigid-body motion. If u and v are the translational components of velocity, I the local image intensity and t, time,

$$uI_x + vI_y + I_t = 0. \tag{1}$$

Horn & Schunck present techniques for solving (1). A substantial modification that I introduced was to estimate a pixel velocity field by using cross correlation first (see figure 8 for the results) and use these estimates as initialization parameters for the Gauss–Seidel iterative solution.

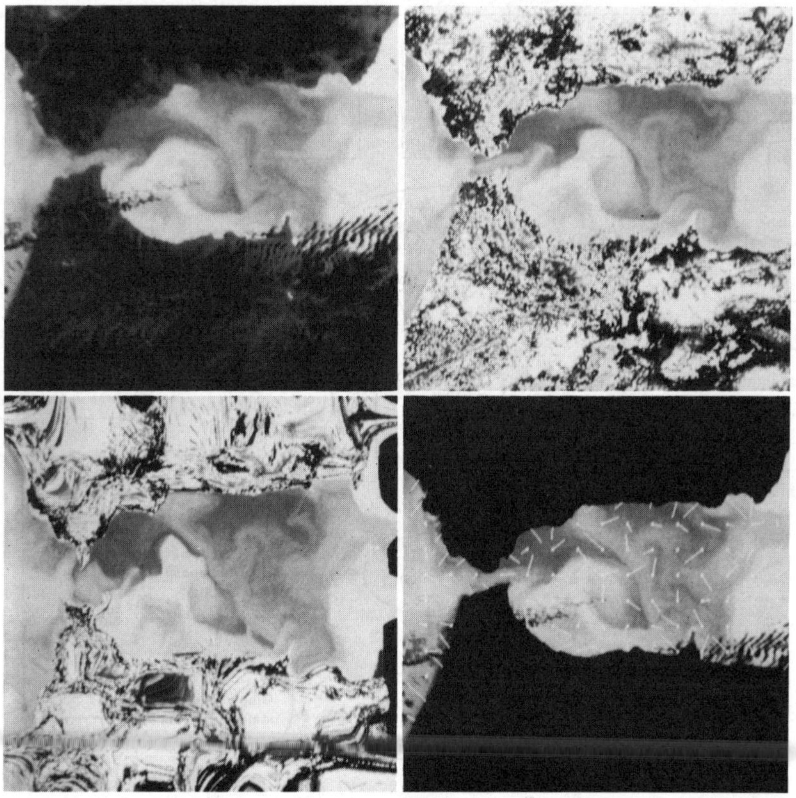

FIGURE 8. Example of optical flow applied to NOAA-AVHRR images. The upper figures show a daytime and night-time image; the lower right shows the velocities extracted automatically, averaged for 32-pixel windows. This type of eulerian velocity is directly compatable for immediate inclusion in numerical simulation. The lower left figure shows an attempt to verify the quality of the output by using the velocities to distort the second image with respect to the first, the resultant being a difference image, indicative of where the technique may be failing.

Equation (1) should be compared with a simple two-dimensional tracer conservation equation (see Stow 1985)

$$uC_x + vC_y - C_t = 0. \tag{2}$$

This similarity in the form of the equations may be used to introduce a process model into the image understanding system, subject to the physical constraints: for optical flow the pixel motion fields must be locally smooth; for tracer conservation, there must be no sources and sinks; and the intensity fields are representative of a conservative passive tracer (in this case, sea-surface temperature). This technique has been successfully applied to time sequences of satellite images of Jupiter from the *Voyager* spacecraft (see Muller 1982), and water-vapour images taken by *Meteosat*.

The second example concerns the automation of the extraction of three-dimensional coordinate information from the *SPOT* (see Muller *et al.* 1988), satellite, being performed within Alvey-funded project MMI/137 on *Real-time 2.5D vision systems*.

Figure 9 shows a schematic of the processing steps involved in extracting surface measurements of terrain (2.5D descriptions) automatically from overlapping *SPOT* images. Inspection of this figure shows a plethora of knowledge sources required: satellite ephemeris (platform parameters in figure 9), digital maps for ground control features (GCF in figure 9) used for

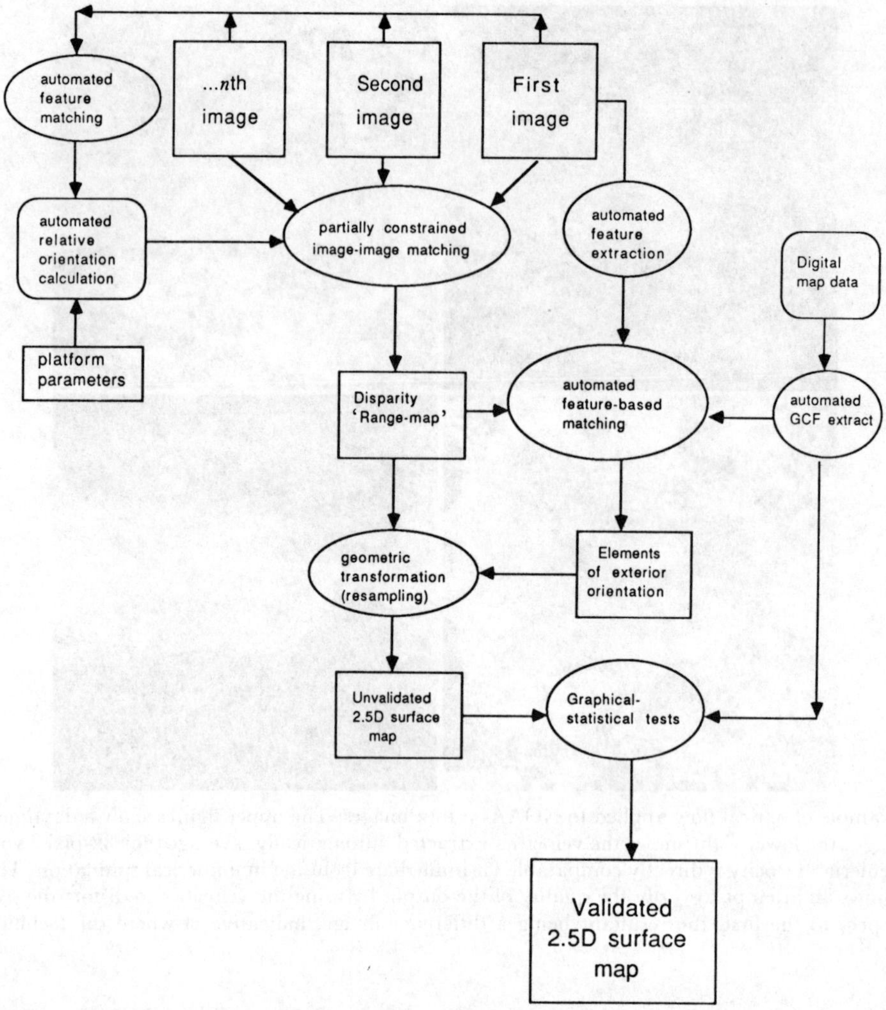

FIGURE 9. Flow-line model of automated stereo matching system for the automated generation of DEMS from *SPOT* images.

determining both the elements of exterior orientation and for validation that DEMS meet the target quality criteria; feature extractors optimized for *SPOT* topographic images and cost functions used for feature-based matching. The system is currently being implemented on a flexible parallel array of transputers (see figure 10) to enable processing of selected windows at video refresh rates.

4. CONCLUSIONS

In this paper, I have tried to indicate the reasons why a new approach to remote sensing is required if we are going to be able to process the flood of data promised by a new generation of satellite sensors (see Goetz *et al.* 1985) into meaningful information that can be incorporated into Geographical Information Systems (see Jackson, this symposium).

The complexity and difficulty of this task cannot be overstated, but current advances in image understanding research indicate which directions appear to be promising. A new era in remote sensing is dawning when quantitative analysis, based on image simulation, will be used

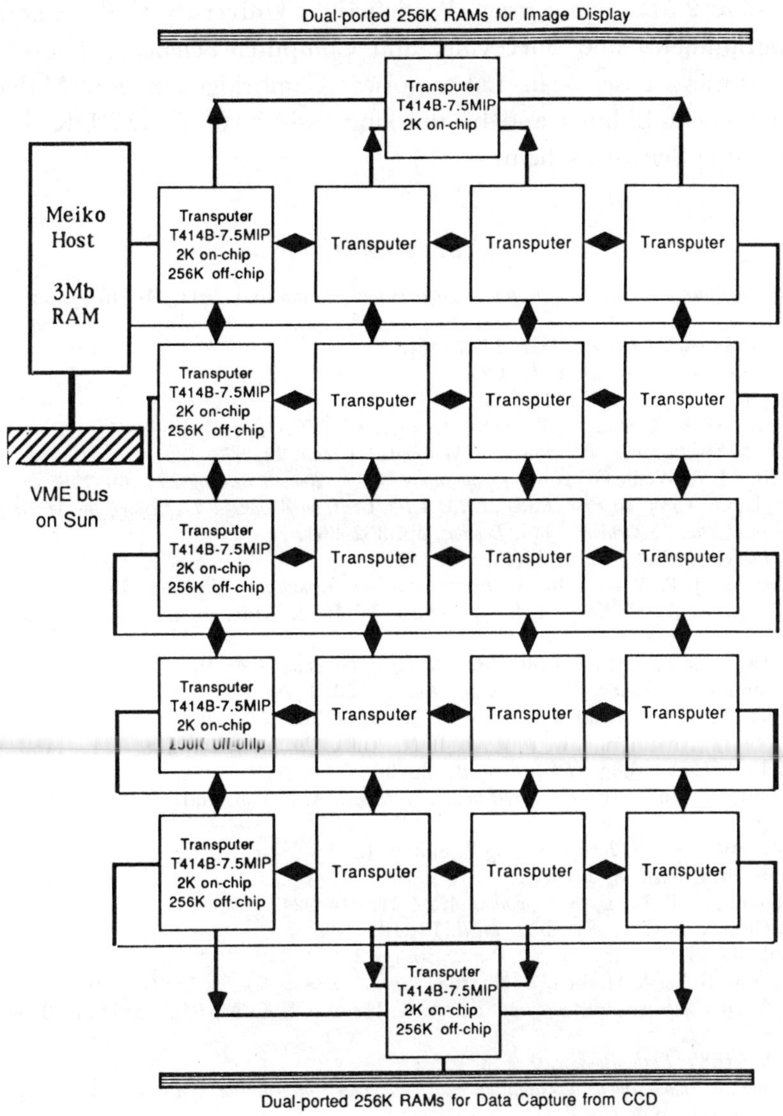

FIGURE 10. Schematic diagram of a transputer array being used at RSRE Malvern in developing a real-time 2.5D vision system for automatically extending DEMs from *SPOT* images.

to extract the information content automatically from satellite images in an analogous fashion to the interplay between theory and empiricism which has led to the successful development of numerical weather prediction.

I thank my students, particularly James Pearson and Philip Eales for providing some of the figures. Thanks also to Michael Dalton, for discussions on ray-tracing; Sam Richards and Tim Day for their work on terrain visualization; Paul Otto for discussions on DEM extraction; Ian Dowman for advice on photogrammetry; Kevin Collins, Brian Roberts, David Watson and Gordon Harp for giving me insight into understanding transputer networks; Mike Jackson, Alan Stevens, Andy Morris and Mark O'Neill for stimulating discussions on map–image matching; and Richard Clarke and Geoff Lawrence for advice on petrogeological applications. The research reported here is partly supported by SERC under the Alvey project number

MMI/137 (*Real-time 2.5D vision systems* (Participants: University College London (Departments of Photogrammetry and Surveying, and Computer Science); Thorn EMI Central Research Laboratories; Laser-Scan Laboratories, Cambridge; RSRE Malvern)); by BP Petroleum Development Limited and by the kind assistance of IBM UK Scientific Centre through their Visiting Scientist scheme.

References

Bernstein, R. 1986 In *Space science and applications: progress and potential* (ed. J. H. McElroy), pp. 71–76. New York: IEEE Press.
Besl, P. & Jain, R. 1985 *ACM Comput. Surv.* **17**, 75–145.
Binford, T. O. 1982 *Int. J. Robot. Res.* **1**, 18–62.
Bretherton, F. P. 1985 *Proc. IEEE* **73**, 1118–1127.
Brown, C. M., Curtiss, M. B. & Sher, D. B. 1983 In *Proc. 8th IJCAI,* pp. 1081–1085.
Cabral, B., Max, N. & Springmeyr, R. 1987 *ACM Comput. Graph,* **21**, 273–282.
Chevrel, M., Courtis, M. & Weill, G. 1981 *Photogramm. Engng Remote Sensing* **27**, 766–778.
Cross, A. & Mason, D. C. 1985 In *Proc. RSS/CERMA Int. Conf. on Advanced Technology for Monitoring and Processing Global Environmental Data, September 1985, London,* pp. 352–364.
Doyle, F. J. 1982 *Int. Arch. Photogrammie* **24**(1), 180–185.
Dowman, I. J. & Muller, J.-P. A. L. 1986 In *Proc. Auto-Carto London, 15–19 September 1986.*
Emery, W. J., Thomas, A. C., Collins, M. J., Crawford, W. R. & Mackas, D. L. 1986 *Eos, Wash.* **67**, 498–499.
Fujimoto, A., Tanaka, T. & Iwata, K. 1986 *IEEE Comput. Graphics Appl.* 16–26.
Gerstl, S. A. W. & Simmer, C. 1986 *Remote Sensing Environ.* **20**, 1–29.
Goetz, A. F. H., Rock, B. N. & Rowan, L. C. 1983 *Econ. Geol.* **78**, 573–590.
Goetz, A. F. H., Vane, G., Solomon, J. E. & Rock, B. N. 1985 *Science, Wash.* **228**, 1147–1153.
Havens, W. & Mackworth, A. 1983 *IEEE Computer* 90–96.
Herman, M. & Kanade, T. 1986 In *From pixels to predicates* (ed. A. P. Pentland), pp. 322–358. New Jersey: Ablex Publishing Corp.
Ho, D. & Assem, A. 1986 *Int. J. Remote Sensing* **7**, 895–904.
Horn, B. K. P. 1977 *Artif. Intell.* **8**, 201–231.
Horn, B. K. P. & Bachman, B. L. 1978 *Communs ACM* **21**, 914–924.
Horn, B. K. P. & Schunck, B. G. 1981 *Artif. Intell.* **17**, 185–199.
Jones, M. & Colombeski, N. C. 1981 *Int. J. Remote Sensing* **2**, 331–342.
Justice, C. O., Wharton, S. W. & Holben, B. N. 1981 *Int. J. Remote Sensing* **2**, 213–230.
Levine, M. D. 1978 In *Computer vision systems* (ed. A. R. Hanson & E. M. Riseman), pp. 335–352. New York: Academic Press.
Maitre, H. & Wu, Y. 1987 *Path. Recog.* **20**, 443–462.
Marr, D. 1982 *Vision: a computational investigation into the human representation and processing of visual information.* San Francisco: W. H. Freeman & Co.
Matsuyama, T. 1987 In *IEEE Trans. Geosci. Rom. Sens.* **GE-25**, 305–316.
McKeown, D. M. Jr., Harvey, W. A. Jr. & McDermott, J. 1985 *IEEE Trans. Pattern Analysis Mach. Intell.* **PAMI-7**, 570–585.
Morgan, J. 1979 *ESA Bull.* **20**, 14–16.
Muller, J.-P. A. L. 1982 Studies of Jovian meteorology using spacecraft and Earth-based images. Ph.D. thesis, University of London.
Muller, J.-P. A. L., Dowman, I. J., Day, T., Dalton, N., Paramananda, V., O'Neill, M. A., Morris, A. C., Stevens, A. & Jackson, M. J. 1987 In *Proc. Remote Sensing Society Conf., Nottingham, 7–11 September 1987.*
Muller, J.-P. A. L. & Day, T. 1987 In *Proc. Joint Photogrammetric and Remote Sensing Society Symposium, London, 11 November 1987.*
Muller, J.-P. A. L., Anthony, A., Brown, A., Chau, K. W., Collins, K. A. C., Dalton, N. M., Day, T., Deacon, A. T., Dowman, I. J., Ibbs, T. J., Jackson, M. J., Jobson, C., Montgomery, P. M., Morris, A. C., O'Neill, M. A., Otto, G. P., Paramananda, V., Recce, M., Robertson, G. W. R., Roberts, J. B. G., Stevens, A., Watson, D. M., Wilbur, S. 1988 First results from Alvey MMI-137. Real time 2.5D Vision Systems. *Int. J. Comput. Vision* (Submitted.)
Nagao, M. & Matsuyama, T. 1980 *A structural analysis of complex aerial photographs.* New York: Plenum Press.
NASA 1979 In *Proc. of the Large Area Crop Inventory Experiment (LACIE) Symposium.* NASA Johnson Space Center, JSC-16015.
Nazif, A. M. & Levine, M. D. 1984 *IEEE Trans. Pattern Analysis Mach. Intell.* **PAMI-6**, 555–577.
Pearce, W. A. 1986 *Appl. Opt.* **25**, 438–447.

Preston, K. Jr. & Duff, M. J. B. 1984 *Modern cellular automata.* New York: Plenum Press.
Reid, C. A., Fung, R. M., Tong, R. M. & Tse, E. 1985 In *Proc. IEEE 2nd Conf. on Artificial Intelligence Applications, 11–13 December 1985,* pp. 125–129.
Rock, B. N. 1984 In *Proc. Int. Symp. on the Remote Sensing of the Environment, 3rd Thematic Conference on Remote Sensing for exploration geology, Colorado Springs, April 16–19, 1984.*
Schachter, B. 1980 In *NATO Workshop on Map Data Processing* (ed. H. Freeman & G. Pieroni). New York: Academic Press.
Slama, C. C. (ed.) 1980 *Manual of photogrammetry,* 4th edn. American Society of Photogrammetry and Remote Sensing.
Stow, D. 1985 *Int. J. Remote Sensing* **6**, 1855–1860.
Swain, P. H. 1985 *Proc. IEEE* **73**, 1031–1039.
Tailor, A., Cross, A., Hogg, D. C. & Mason, D. C. 1986 *Image Vision Comput.* **4**, 67–83.
Teillet, P. M. 1986 *Int. J. Remote Sensing* **7**, 1637–1651.
Tenenbaum, J. M., Barrow, H. G., Bolles, R. C., Fischler, M. A. & Wolf, H. C. 1979 In *Proc. IEEE Workshop on Pattern Recognition in Image Processing, 6–8 August 1979, Chicago,* pp. 610–617.
Tenenbaum, J. M., Fischler, M. A. & Barrow, H. G. 1980 *Comput. Graph. Image Process.* **12**, 407–425.
Woodham, R. J. & Lee, T. K. 1985 *Can. J. Remote Sensing* **11**, 132–161.

Discussion

J. C. Scott (*A.R.E., Portland, Dorset, U.K.*). In the context of the data continuously available to the human being from his senses, the 2% quoted as processed from some satellites actually seems quite high. It is important to separate the customers and users of remote sensing into categories: those who need continuous global coverage at one end of the scale, and those needing occasional small regions at the other; immediacy of data is another factor. Would Dr Muller comment on how these factors affects the division between human interaction and automation?

J.-P. A. L. Muller. In the short term, human interaction will continue to play a key role in data analysis, even when guided by automated processing. An example of this that I showed was the visualization of terrain that could incorporate automatically generated DEMS.

In the longer term, human interaction will be involved in quality control of the output of an image-understanding system, both as it reaches intermediate conclusions and for final results.

S. Quegan (*Department of Applied and Computational Mathematics, The University, Sheffield, U.K.*). From the various examples quoted, can Dr Muller recognize any common elements, or does each problem have to be treated entirely in its own right?

J.-P. A. L. Muller. Common elements consist of the role of image-formation models, the multilevel invocation of multifarious knowledge sources and the matching processes needed for object–scene recognition. However, the type, quantity and quality of each knowledge source and when and where it is invoked is specific to a specialist domain application.

Phil Trans. R. Soc. Lond. A **324**, 397–407 (1988)
Printed in Great Britain

The role of prior knowledge in coherent image processing

BY S. P. LUTTRELL AND C. J. OLIVER

Royal Signals and Radar Establishment, St Andrews Road, Great Malvern, Worcester WR14 3PS, U.K.

[Plates 1 and 2]

Models that encode prior knowledge about a scene provide a means for interpreting image data from that scene in more detail than would otherwise be so. Information about both background clutter and target characteristics should be included in this prior knowledge. We demonstrate the use of a generalized noise model to represent a variety of naturally occurring random terrain clutter textures observed in high-resolution synthetic aperture radar (SAR) images. In addition a similar approach is adopted for the simulation of such textures. Having established the background properties we next introduce prior knowledge about any target within the scene and exploit this in achieving a cross-section reconstruction having improved resolution compared with the original image. Examples of such a super-resolution method based on singular value decomposition are demonstrated and the limits of the technique are indicated.

INTRODUCTION

In this paper we shall be concerned with aspects of the interpretation of synthetic aperture radar (SAR) images. In common with all image interpretation processes we shall imbue the raw-image data with some 'meaning' based on relevant past experience in a similar context. This 'prior knowledge' is cast in the form of a model that underlies our image interpretation.

In many radar applications we are concerned with detecting the presence of 'targets' against background 'clutter'. This statement assumes more information about the scene beyond that found in the image alone. We have introduced prior knowledge that (typically) targets are small bright blobs embedded in a weak clutter background. To detect the presence of such targets we must develop an understanding of the properties of both targets and clutter. In this paper we describe techniques to increase our understanding of both classes by the application of prior knowledge.

CLUTTER MODELS AS PRIOR KNOWLEDGE

In figure 1, plate 1, we show a SAR image of a region of fields and woods, obtained with the RSRE X-band airborne system. It is apparent that the intensity in the image shows strong speckle behaviour. This is a well-known phenomenon in coherent imaging and arises where the total intensity is made up by the interference between large numbers of randomly positioned scatterers within each resolution cell. The presence of this speckle over the entire image suggests that a *physical model* for scattering and imaging of the surface might describe each resolution cell as containing many randomly positioned scatterers. No attempt is made to represent the configuration within the cell. The physical processes of scattering and imaging can then be analysed theoretically. Where the cross section is spatially non-varying the image statistics should correspond to classical Rayleigh speckle. In table 1 we compare the higher-order single-point moments of the detected intensity distribution for a region of open field, for which the cross section is expected to be essentially constant, with the theoretical values for pure speckle.

TABLE 1. COMPARISON OF THE INTENSITY MOMENTS FOR THE FIELD TEXTURE

		model	
moment	data	mean	SD
second	1.970	2.000	0.014
third	5.68	6.00	0.14
fourth	21.3	24.0	1.3

It is apparent that the statistics are indeed consistent with such a model. In general the detected intensity of each pixel of an image will then take the form of a convolution of the underlying cross-section fluctuation from pixel to pixel with a negative exponential speckle contribution. This process is clearly visible qualitatively in figure 1. One would therefore expect that the intensity probability distribution will no longer be negative exponential but will have larger higher-order moments indicating the presence of excess fluctuations.

Having proposed a physical scattering model consistent with the observed speckle it is next necessary to provide a model for the cross-section fluctuations. In general, no physical description of the surface appears feasible so we are forced to consider a *phenomenological model* for the surface cross-section fluctuations. The aim of such a model is to represent the surface in such a way that the scattered radiation is consistent with the properties of the observed image; there is no attempt to reproduce the exact surface. We shall show that natural clutter textures can be treated as homogenous noise processes. In terms of figure 1, therefore, we are concerned that the model should be capable of reproducing the statistical and correlation properties of clutter textures such as that shown enclosed by the white rectangle.

One such model that has been proposed is that the surface cross section, σ, should be represented by a random process with a gamma-distributed probability-density function (PDF)

$$p(\sigma) = \frac{1}{\langle\sigma\rangle}\left(\frac{\sigma}{\langle\sigma\rangle}\right)^{\nu-1}\frac{1}{\Gamma(\nu)}\exp\left[-\frac{\sigma}{\langle\sigma\rangle}\right], \tag{1}$$

where $\Gamma(\nu)$ is the gamma function of order ν (Jakeman & Pusey 1976, 1977, 1978; Jakeman 1980*a*). This is the continuum analogue of the discrete negative-binomial distribution. On scattering the observed intensity PDF would then be the convolution of the gamma-distribution with the negative exponential distribution, namely the *K*-distribution given by

$$p(I) = \frac{1}{\langle I\rangle}\left(\frac{I}{\langle I\rangle}\right)^{\frac{1}{2}(\nu-1)}\frac{1}{\Gamma(\nu)}K_{\nu-1}\left[2\left(\frac{I}{\langle I\rangle}\right)^{\frac{1}{2}}\right], \tag{2}$$

where $K_{\nu-1}$ is the modified Bessel function (*K* distribution) of order $\nu-1$. From (1) it is apparent that this noise model is a generalization of thermal noise (the case when $\nu = 1$). The order parameter ν serves to control the contrast and higher-order moments of the distribution. Such a noise model would result from a physical model in which the local cross section, or density of scatterers, depended on a birth–death-migration process (Jakeman 1980*b*). Thermal noise describes a process in which only migration (diffusion) occurs. This noise process is the result of including an additional source and sink of scatterers. A variety of evidence from scattering experiments at both microwave and optical wavelengths has indicated that such a model is

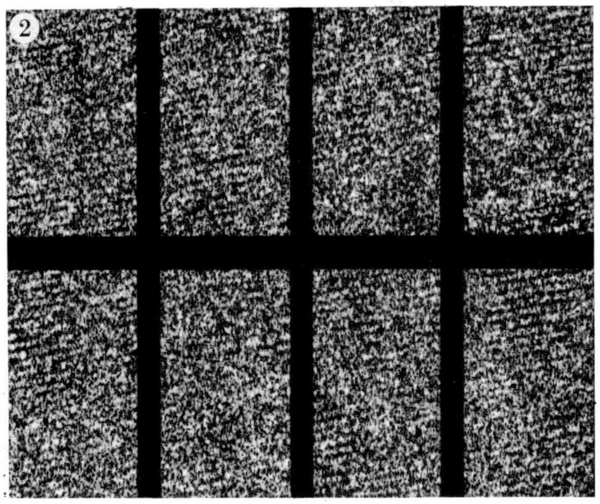

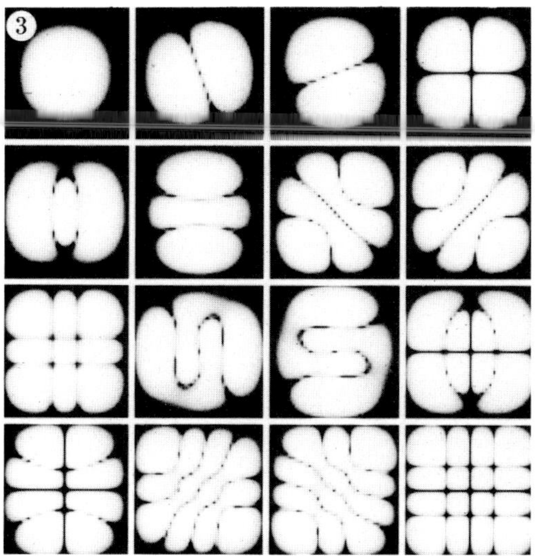

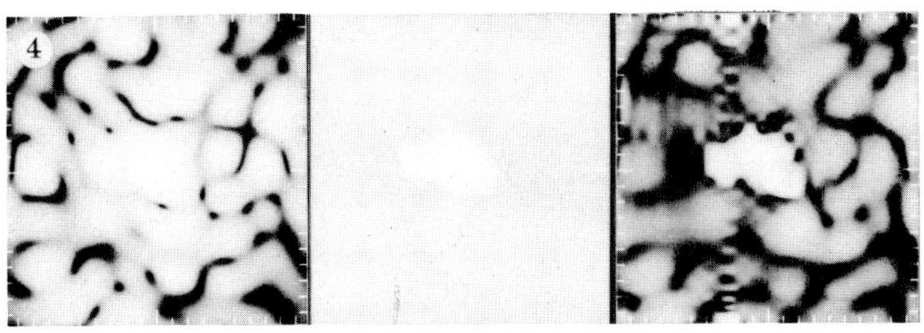

FIGURE 1. SAR image of a region of fields and woods.

FIGURE 2. Comparison of original (top right) and several simulated textures for the region indicated in figure 1.

FIGURE 3. First 16 of the weighted object eigenvectors when no prior knowledge is introduced; logarithmic display.

FIGURE 4. Stages in the first iteration of the reconstruction process for the simulated target. Original image (left), prior knowledge weight (centre), reconstruction (right); logarithmic display.

(*Facing p.* 398)

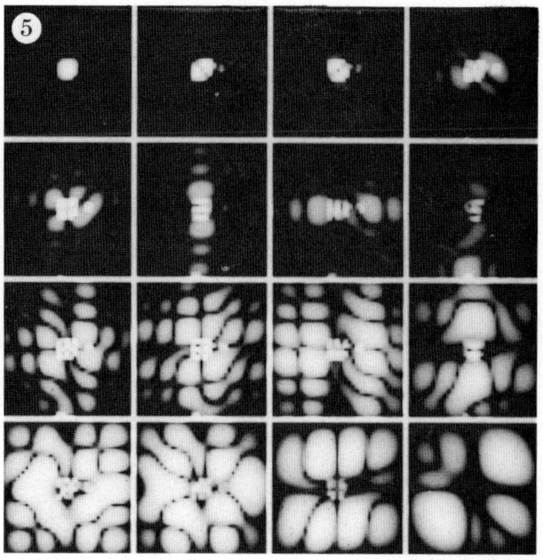

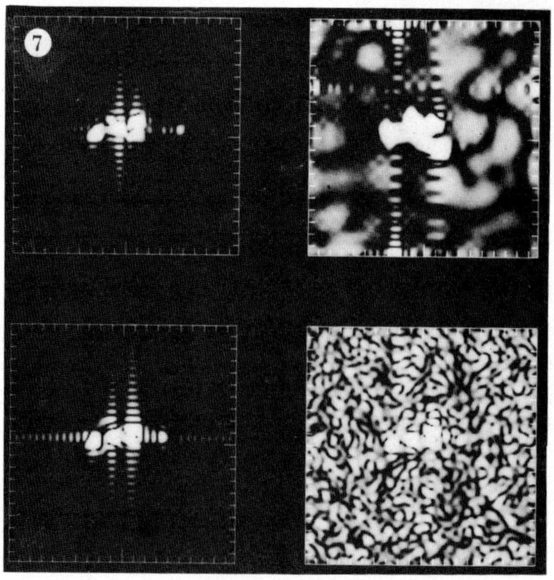

FIGURE 5. First 16 of the weighted object eigenvectors when localized target prior knowledge is introduced; logarithmic display.

FIGURE 6. Stages in the second iteration of the reconstruction process for the simulated target. First reconstruction (left), prior knowledge weight (centre), reconstruction (right); logarithmic display.

FIGURE 7. Comparison of the final reconstructions with high resolution images of the same object. Reconstruction without background (top left), reconstruction with background (top right), high-resolution image without background (bottom left), high-resolution image with background (bottom right); logarithmic display.

consistent with the data (Jakeman & Pusey 1975; Jakeman *et al.* 1976; Parry *et al.* 1977; Ward 1981, 1982; Oliver 1984, 1986*a*, *b*, 1988) though no unambiguous physical justification is possible.

Before embarking on a detailed comparison of the statistics it is essential to consider the correlation properties within the texture which often convey more information than the contrast fluctuations. Gamma-distributed noise can be regarded as the resultant of a random walk in ν dimensions of a complex gaussian variable, i.e. thermal noise. Thus all the results for classical thermal noise can be generalized to the gamma-distributed equivalent by exploiting the appropriate factorization properties (Oliver 1984). For distributions with high contrast it is necessary to use fractional values of the order parameter ν. Provided that the model truly represents the texture, it then contains fundamentally all the information available about that texture because gaussian processes are described entirely in terms of their mean and covariance. Various simple theoretical models of this kind have been proposed and compared with image data (Oliver 1984, 1985, 1986*b*, 1988).

Let us next apply this approach to the texture of the plantation of young trees shown in figure 1 enclosed by a white rectangle. The first stage is to derive a reasonable model for the surface autocorrelation function (ACF) which, when imaged, gives an intensity ACF that closely represents the observed image result. For the texture shown the model requires three components for an approximate correspondence. A method also exists for obtaining even closer agreement by regarding the ACF of the original data as a perturbation of the model ACF (Oliver 1987). Having obtained the form of the surface ACF the higher-order moments can then be calculated following the methods established earlier (Oliver 1984, 1985, 1986*b*). A comparison of the observed single-point moments with the predicted values for the first four moments is given in table 2. In addition the predicted values for a simple K-distribution having the observed second moment are included.

TABLE 2. COMPARISON OF THE INTENSITY MOMENTS FOR THE WOODED TEXTURE DENOTED IN FIGURE 1

		full model		simple model		simulation	
moment	data	mean	SD	mean	SD	mean	SD
second	2.64	2.64	0.04	2.64	0.06	2.55	0.11
third	13.1	13.1	0.7	13.0	1.4	14	3
fourth	100	108	10	102	32	160	90

Both theoretical predictions are consistent with the original data within the theoretical errors included in the table. Strictly speaking an integrated K-distribution is not itself K-distributed so the simple model should not fit as well. However, for distributions with lower contrast than thermal noise the difference is often insignificant. The simple model, of course, is not capable of predicting how the statistics vary with the imaging resolution; for this the full theory is required.

We have so far demonstrated that random textures in coherent images can be represented in terms of a correlated gamma-distributed surface cross section, which when imaged yields K-distributed correlated intensity fluctuations. Although only a single example is included here many other clutter textures in both radar and sonar images have been demonstrated to be capable of representation with the same fundamental model.

As a final test of the model it is possible to simulate textures having the same statistical properties as the original image by using a linear filter method (Oliver & Tough 1986; Oliver 1986*b*). This method seeks to reproduce the two-point statistics, the covariance function, of the original image. The surface is first simulated by passing uncorrelated gamma-distributed noise through a linear filter. The filter weights are chosen so that their ACF is determined by the surface ACF (Oliver & Tough 1986). The complex scattered field is then simulated by speckling the cross section with a random zero-mean gaussian variable for both the in-phase (I) and quadrature (Q) field components. This field is then convolved with the imaging response function of the system. A set of such simulations is compared with the original texture (top right) in figure 2, plate 1. Visually the agreement is excellent. As a further test the mean value and standard deviation of the higher-order single-point statistics of the simulations is included in the last two columns of table 2. Again the agreement is within the statistics. This ability to simulate textures is of great importance in deriving statistical estimates of, for example, target detection probabilities obtained with various algorithms against such a clutter background.

The highest level of prior knowledge, the *contextual model*, describes the relation of defined textures within a scene. In terms of figure 1, therefore, it would contain map information on the position of the boundaries of the woods and fields, statistical information about the correlation properties and order parameter of the texture in each region, plus other information, such as tree height, ground slope and aircraft altitude, to enable one to predict shadowing and other geometrical effects. In principle, the combination of all three stages in the model should enable us to predict coherent images that are statistically equivalent to a SAR image of the same region.

Once the model for the clutter is established it can be used in the detection of targets against this clutter background. The first stage in the process is to form a composite clutter model for the area created from combinations of map data, previous SAR images of the same area and other sensor data. This clutter model (prior knowledge) then defines the predicted statistical properties that are expected at each point in the SAR image. To detect 'targets' we then start with the assumption that the scene is entirely described in terms of the clutter map and search for anomalies which we define as targets. These anomalies may not be simply related to the image intensity but may depend on higher-order statistical properties of the image. We might, for instance, wish to detect changes of texture correlation length where the mean intensity and contrast remain the same. Whatever criterion is selected for the discrimination the clutter model is used to define a 'target detection threshold map' for that measure. Target detection then occurs when the value of the measure exceeds this threshold.

Target resolution-enhancement from prior knowledge

Once a target has been detected the next stage is to seek to classify that target. In many examples of SAR image the resolution of the system is of the same order as the size of targets of interest making any type of geometrical classification based on the distribution of cross section within the target essentially impossible. The second part of this paper now concentrates on the use of prior knowledge about targets to improve the effective resolution of the sensor. This is achieved by introducing a model to represent the target. There are therefore two key issues that have to be considered: (i) what is an appropriate form for the prior knowledge model and (ii) how is the model applied to yield the maximum additional information about the target?

Before addressing these issues directly let us consider a particular example of a prior-knowledge model and examine the implications. The Rayleigh resolution criterion states that two point sources are just resolved when the first null of one point spread function (PSF) lies on the main lobe of the second. However, in the 'monopulse' technique of radar processing strong prior knowledge about the target is introduced, namely that it consists of two scatterers. We can then inspect the detailed shape of the image to determine the amplitudes, phases and positions of the two scatterers with greatly enhanced resolution, limited only by the noise in the system, not the PSF directly. This type of approach is common in many experimental fields where one defines a model to represent the process under study and then determines the parameters of that model. The approach is effective provided that the prior knowledge introduced in the model is in fact correct. It is therefore advisable to use as weak a form of prior knowledge as possible because there is less chance of introducing false information. Let us now consider what general properties of targets might be appropriate as prior knowledge in the context of SAR images.

Obviously it would be possible to generalize the monopulse technique by assuming that one had a number of point sources within the target and searching for the best fit as that number is varied. As the number increases the ambiguity introduced by the noise means that a large number of alternative solutions exist and one is forced to introduce some arbitrary criterion for choosing a particular solution, for example, that with the smallest number of scatterers. This form of prior knowledge is generally speaking too strong because we cannot be sure that targets consist of point-like scatterers. A weaker form of prior knowledge that we can be confident about is that targets are of limited spatial extent. Furthermore we may assert that the target's position and dimensions are well represented by the original image. The fundamental premise of all radar target detection algorithms is that a target exists where a large return is recorded. This form of prior knowledge is based on the same assumption and may be viewed with the same level of confidence. It is obviously true that pathological scatterer configurations can violate the assumption and so invalidate the result. One could only have complete confidence in the prior knowledge if one knew precisely what the target was, in which case there is no more information to be gained from the image. This concentration of the target cross section into a localized continuum form makes no assertions about the scattering centres within the target itself; this information will be derived by applying the prior knowledge to the image data.

In applying resolution-enhancement methods on an SAR image a small region enclosing a detected target will be selected. Thus the prior knowledge must combine background and target models. For the total model we assert that it is made up of a background region of uniform cross section containing many random scatterers per resolution cell. Imposed on this background at a position and with dimensions determined from the original image is a localized target within which the scatterer distribution is unknown. Comparative strengths of target and background can be deduced from the image data.

Let us now examine how this prior knowledge may be exploited in the reconstruction of a target at an enhanced resolution. We shall follow the methods previously proposed by Luttrell (1985*a*) and Luttrell & Oliver (1986) based on a bayesian reconstruction scheme. Suppose that our object function, f ($=f(x)$, a continuous complex variable denoting the scattering amplitude and phase), is imaged through a response function T to give a complex image vector $\boldsymbol{g}$. The imaging process can be denoted by the shorthand notation

$$\boldsymbol{g} = Tf+\boldsymbol{n}, \tag{3}$$

where the vector $\boldsymbol{n}$ is the noise introduced in the process. Bayesian reconstruction then provides a means for relating the *a posteriori* PDF, $P[f(x)\,|\,\boldsymbol{g}]$ (the conditional probability of a particular object given the image data), to the prior knowledge of the object, expressed in terms of the *a priori* PDF $P[f(x)]$ (the predicted probability of a particular object) and the imaging process via the expression

$$P[f(x)\,|\,\boldsymbol{g}] \propto P[\boldsymbol{g}\,|\,f(x)]\,P[f(x)], \tag{4}$$

where $P[\boldsymbol{g}\,|\,f(x)]$ describes the imaging properties in terms of the conditional probability of the image given a particular object. Note that a one-dimensional form is used for simplicity. This *a posteriori* PDF is the basic construct from which inferences about the data, in particular the target reconstruction, are to be drawn.

To progress we need to make assumptions about the PDFs of the object and the noise. Let us assume that the object distribution, the *a priori* PDF, is a zero-mean gaussian. This will not accurately represent the non-gaussian statistics that actually occur but will suffice to specify the scale of the amplitude of $f(x)$. In principle we cannot specify what the statistics of the object should be without knowing in advance what the target is. Thus the gaussian assumption is a reasonable compromise. The properties of the object are then all contained in its covariance matrix, W. In the imaging process in the absence of noise the conditional PDF $P[\boldsymbol{g}\,|\,f(x)]$ will be a delta function because the mapping is determined precisely by the imaging function T. When noise is introduced it broadens the conditional PDF. Let us assume that the measurement noise is also a zero-mean gaussian process specified completely in terms of its covariance matrix, N. Note that it is not necessary to assume that the noise is uncorrelated; correlated noise can be treated correctly. It can then be shown (Luttrell 1985*a*; Luttrell & Oliver 1986) that the *a posteriori* PDF can be expressed in the form

$$P[f(x)\,|\,\boldsymbol{g}] \propto \exp\left[-(f-\langle f\rangle)^{\dagger}\,(T^{\dagger}N^{-1}T+W^{-1})\,(f-\langle f\rangle)\right], \tag{5}$$

where $\langle f\rangle = WT^{\dagger}(TWT^{\dagger}+N)^{-1}\boldsymbol{g}$ († denotes hermitian conjugate). (6)

This *a posteriori* PDF contains the available information about the object. In practice we require to select some particular member of this distribution as representing the 'best' reconstruction. A logical choice would be to choose the average value of $f(x)$, defined in (6).

Let us now make the simplifying assumptions that both the object and noise covariances are delta-correlated. The 'weight' W is now a real function which can be related to a profile function $P(x)$ such that

$$W(x_1, x_2) = |P(x_1)|^2\delta(x_1-x_2). \tag{7}$$

It can be shown (Luttrell 1985*a*; Luttrell & Oliver 1986) that the object reconstruction may then be expressed as

$$f_{\text{rec}}(x) = P(x)\sum_{k=1}^{m}\frac{g_k}{(\lambda_k+N)^{\frac{1}{2}}}u^k(x), \tag{8}$$

where g_k are the components of the image $\boldsymbol{g}$ in the image basis vectors and $u^k(x)$ are the equivalent eigenvectors in object space. These object and image eigenvectors correspond to singular value decomposition (SVD) of the imaging process and are characterized by the eigenvalues λ_k. If the signal to noise ratio is large enough the results of this bayesian reconstruction method are identical to those obtained elsewhere using an SVD technique (Bertero & Pike 1982; Bertero *et al.* 1984*a, b*).

Let us next examine the *a posteriori* distribution in more detail to determine where the dominant information is contained. It can be shown that any object can be expanded in terms of object space vectors in the form

$$f(x) = P(x) \sum_{k=1}^{m} \frac{f_k}{(\lambda_k + N)^{\frac{1}{2}}} u^k(x) + f_c(x), \tag{9}$$

where the expansion coefficients f_k are the components of the image of the object f in the image space. The second term in (9), f_c, denotes the sum of the components of f that do not project onto the image space basis. Because these components have no contribution in the image (data) space they can be said to be 'invisible' to the data. Thus we cannot infer anything about these components from a measurement of the image $\mathbf{g}$. All that we can say about them is contained in the form of the *a priori* PDF that we have defined.

If we next consider the implication of the first term in (9) we see that the contribution to the reconstructed object of each of the vectors f_k is characterized by an eigenvalue λ_k, which defines the visibility of that contribution. Where these eigenvalues are smaller than the noise N the corresponding data have insufficient energy to dominate the noise and so cannot convey information about the object reconstruction. In this situation also the only information available is that introduced in the prior knowledge. Only those contributions f_k whose eigenvalues exceed the noise provide additional information beyond that represented by the prior knowledge.

A more detailed discussion of the information content of these images, expressed in terms of mutual information, is to be found in Luttrell (1985*b*, *c*) and Luttrell & Oliver (1986).

If we next substitute for f and $\langle f \rangle$ ($=f_{\text{rec}}$) from (8) and (9) into (5) we obtain for the *a posteriori* PDF

$$P[f(x) \,|\, \mathbf{g}] \propto \exp\,[f_c^+ |P|^{-2} f_c]\, \exp\left[-\sum_{k=1}^{m} |f_k - g_k|^2 \left(\frac{1}{N} - \frac{1}{\lambda_k + N}\right)\right]. \tag{10}$$

The first factor corresponds to a piece of the *a priori* PDF that is not changed in passing to the *a posteriori* PDF, since f_c is invisible in the data space. The second factor depends on the data and expresses the influence that the data, $\mathbf{g}$, have on the *a posteriori* PDF. Each mode carries an independent piece of useful information about the object, which vanishes as the noise increases.

Let us now illustrate the process by using (8) to calculate reconstructed objects for a simulated target embedded in clutter. We shall simulate a two-dimensional imaging system whose PSF is that for a square aperture. The data acquisition system measures complex field Nyquist-spaced samples (I and Q) on a square lattice. Because the target is localized we shall consider an image space of 9×9 samples. This must be sufficiently large that no sidelobes of the target are visible outside the sampled image. In the object space each image sample is subdivided into 4×4 subsamples without lying outside the bounds of the 9×9 array. This sampling is deemed adequate to provide at least Nyquist sampling on the enhanced-resolution object reconstructions. For the sake of this demonstration a typical small target is simulated which leads to the observed image shown on the left of figure 4, plate 1.

Having defined the sampling spaces we next require estimates of W and N. In typical SAR images the measurement noise and the background clutter speckle are both essentially delta-correlated. Furthermore the noise is negligible compared with the clutter speckle fluctuations and may be ignored. If we also assume that the target is delta-correlated then the target and clutter prior knowledge are expressed in terms of the profile weighting function $W(x) = |P(x)|^2$

which has the form of a cross section. Having determined the weighting function we next compute the reconstructed object $f_{rec}(x)$ from the image $\boldsymbol{g}$ as given in (8). In the analysis it can be shown (Luttrell 1985*a*) that the reconstruction vectors in the object space include the profile function P. It is instructive to examine the consequence of introducing no prior knowledge into the reconstruction. This is equivalent to setting P constant; we are totally non-committal about where the scatterers are located within the image. In figure 3, plate 1, we show the first 16 of the 81 possible object space eigenvectors arranged in order of decreasing eigenvalues. Note that as the eigenvalue decreases the spatial frequency content of the eigenvectors increases. In general higher spatial frequency terms are more attenuated by the imaging process leading to the ordering of the functions. However, for a uniform weighting function all eigenvalues are approximately equal so that all 81 are required to perform the reconstruction. As one might expect, a reconstruction based on this prior knowledge is essentially the same as the original image and can be shown (Luttrell 1985*a*) to provide an interpolated form of the data. We may use this interpolated form to make cross-section estimates for the combined clutter and target prior-knowledge weighting function.

Initially we use threshold detection to separate those pixels that may be defined as 'target' from those we define as 'clutter'. The mean intensity of those pixels that lie below the threshold is taken as the background cross-section estimator whereas the intensity of the pixels that exceed the threshold is taken as the estimate of the local target cross section. This process whereby the data is used to generate 'prior knowledge' may seem incestuous. It is, however, important to appreciate that the prior knowledge does not reside in the image data but in the rules that are used in separating the image into the two components and setting up the weighting function. In particular we are asserting, in common with nearly all other radar systems, that locally bright pixels are extremely likely to have been derived from local concentrations of cross section rather than from the myriad of other possible configurations that would give the identical image. This rule is not a fundamental constraint so much as a consequence of empirical observation of typical SAR images.

Let us now apply these methods to a particular simulated image, shown on the left of figure 4, plate 1. This image has been formed from an array of four dominant scatterers. However, the resolution is so poor that the image shows merely some evidence of asymmetry. The signal to background ratio is 20 dB. Performing threshold detection we derive the prior knowledge cross-section weighting function shown in the centre of the figure. Because the method uses threshold detection there is a statistical probability that pixels which might actually belong to the background class will emerge as targets, and vice versa. Two examples of the former effect can be observed. This weighting function now takes the form of the *a priori* PDF in the bayesian reconstruction scheme. The resulting reconstructed object is shown on the right of figure 4. Comparison with the original image reveals that the apparent resolution in the image has indeed been improved. The degree of enhancement depends strongly on the nature of the data and it is not possible to define a global enhancement factor. Some understanding of how the enhancement is achieved may be gained from studying the object space eigenvectors for the case where a localized target leads to a strongly peaked weighting function. The first 16 of these modes are shown in figure 5, plate 2. Again they are presented in order of decreasing eigenvalue. Note how the first few modes concentrate almost exclusively on a strongly localized region contained within the central peak of the data. No such effect was observed with the uniform weight (figure 3). Unlike the uniform example also, the eigenvalues of the modes decrease comparatively rapidly as the order increases. Thus a small region of the object space contains a large amount of the information present in the data. Only a comparatively small amount of the information, contained in the weak higher-order modes, is used to reconstruct the background clutter. This is the means whereby resolution enhancement is achieved in this approach.

Having applied the resolution-enhancement technique once it is tempting to attempt to apply it again. There are two main objections to such a step. Firstly, one has to vindicate the choice of the prior knowledge in an iterative procedure. Regardless of the true nature of the target this type of processing will tend to concentrate the object cross section into a number of localized scatterers. If the image were of an object such as a tree then this progressive concentration of the cross section would be inappropriate since the scatterers are distributed in a comparatively diffuse manner. On the other hand, man-made objects, such as vehicles, can usually be represented in terms of a small number of point scatterers. Thus, if it were possible to introduce further prior knowledge that the object in question is indeed man-made, we may proceed with further iterations of the reconstruction process treating the output of the first reconstruction as if it were itself an image.

However, proceeding with iterated reconstruction introduces a second limiting factor into consideration. The presence of the random clutter speckle within the image introduces a noise component into the reconstruction so that, eventually, the result is dominated by the noise. It is clearly essential to iterate the process only for as long as this noise component permits. In this instance a second iteration can be justified. Exactly the same procedure is adopted as before with the previous reconstruction treated as the new image (figure 6, plate 2, left).

The weighting function, shown in the centre of figure 6, plate 2, is derived in the same way as before. Note that the threshold detection algorithm still passes the two false alarm targets detected in the original data. The reconstruction shows still more resolution enhancement and even serves to split the original image into four dominant scatterers (figure 6, right).

Let us conclude by comparing the reconstruction with an image of the original object where the resolution is increased by a factor of four in each direction. In figure 7, plate 2, reconstructed images after two iterations with (top right) and without (top left) a clutter background are compared with equivalent high-resolution images (bottom).

The example on the left without background gives a very high-quality reconstruction, closely similar to the image, because the only limiting factor is the noise introduced by rounding errors in the computer. The example on the right, with an initial target to background ratio of 20 dB, shows the distortion that has been introduced by the random speckle. Nevertheless the information content of the reconstruction has been considerably enhanced by the introduction of prior knowledge when compared with the original image (top left, figure 4). Note also that the process has suppressed the background clutter compared to the peaks of the reconstructed target. Although this reconstruction may not in itself be sufficient to classify the target it does represent a significant step in that direction. This has been achieved entirely by post-processing the original data and does not require any modification of the existing SAR sensor.

Conclusions

The task of interpreting SAR images is very difficult. In addition to the problems associated with the complexity of the image and the sheer quantity of data, which it shares with many other sensor systems, there are the further limitations of speckle and low resolution. It would be extremely useful to be able to summarize the content of a SAR image in terms of a few significant features, 'targets', set against a 'clutter' background. One can then concentrate one's efforts on the targets. In the first main section of the paper we presented a model of SAR terrain that could be used to define a target detection threshold and to suggest suitable detection algorithms. Ensuing target detections, which are essentially a result of the first stage of the reconstruction process, can then be studied in more detail including, in particular, resolution enhancement as outlined in the second main section of the paper. Both steps are vital tools in the total problem of SAR image interpretation.

References

Bertero, M. & Pike, E. R. 1982 *Opt. Acta* **29**, 727–746.
Bertero, M., Brianzi, P., Parker, P. & Pike, E. R. 1984*a* *Opt. Acta* **31**, 181–201.
Bertero, M., De Mol, C., Pike, E. R. & Walker, J. G. 1984*b* *Opt. Acta* **31**, 923–946.
Jakeman, E. 1980*a* *J. Phys.* A **13**, 31–48.
Jakeman, E. 1980*b* *Opt. Acta* **27**, 735–741.
Jakeman, E., Pike, E. R. & Pusey, P. N. 1976 *Nature, Lond.* **263**, 215–217.
Jakeman, E. & Pusey, P. N. 1975 *J. Phys.* A **8**, 369–410.
Jakeman, E. & Pusey, P. N. 1976 *IEEE Trans. Antennas Propag.* **AP-24**, 806–814.
Jakeman, E. & Pusey, P. N. 1977 *Radar 77 (IEEE Conf. Publ.* 155), pp. 105–109.
Jakeman, E. & Pusey, P. N. 1978 *Phys. Rev. Lett.* **40**, 546–550.
Luttrell, S. P. 1985*a* *Opt. Acta* **32**, 701–714.
Luttrell, S. P. 1985*b* *Opt. Acta* **32**, 255–257.
Luttrell, S. P. 1985*c* *Inv. Problems* **1**, 199–218.
Luttrell, S. P. & Oliver, C. J. 1986 *J. Phys.* D. **19**, 333–356.
Oliver, C. J. 1984 *Opt. Acta* **31**, 701–722.
Oliver, C. J. 1985 *Opt. Acta* **32**, 1515–1547.
Oliver, C. J. 1986*a* In *Wave propagation and scattering* (ed. B. J. Uscinski), pp. 155–174. Oxford: Clarendon Press.
Oliver, C. J. 1986*b* *Inv. Problems* **2**, 481–518.
Oliver, C. J. 1988 In *Proc. IMA Conf. on Applications of Mathematics to Remote Sensing*. Oxford University Press. (In the press.)
Oliver, C. J. 1987 *Inv. Problems.* **3**, 463–475.
Oliver, C. J. & Tough, R. J. A. 1986 *Opt. Acta* **33**, 223–250.
Parry, G., Pusey, P. N., Jakeman, E. & McWhirter, J. G. 1977 *Opt. Commun.* **22**, 195–201.
Ward, K. D. 1981 *Electron. Lett.* **17**, 561–565.
Ward, K. D. 1982 *Radar 82* (*IEE Conf. Publ.* 216), pp. 203–207.

Discussion

P. L. C. JEYNES (*Oxford Computer Services Ltd*, 244 *Barnes Road*, *Cowley*, *Oxford*). I would have thought that in general the concept of 'individual scatterers' is rather meaningless in wave scattering from rough surfaces.

C. J. OLIVER. Firstly, the concept of a surface is not really applicable for describing radar scattering from the terrain. Secondly, we could not solve scattering from such a surface even if we could represent it. Accordingly, as described in the paper, we adopt a phenomenological model which reproduces, as far as possible, the observed image texture. This model enables us to make assertions about image properties as the operating conditions change and has a reasonable, though non-rigorous, physical interpretation.

S. QUEGAN (*Department of Applied and Computational Mathematics*, *The University*, *Sheffield*). As a proviso to Dr Oliver's last reply, texture described by the autocorrelation function is present only in some land cover types. It is not present in, for example, arable crops, and for these cover types the only information available is the mean backscatter.

J. A. ALLAN (*School of Oriental and African Studies, University of London, U.K.*). For those of us interested in 'backgrounds' rather than in objects disturbing backgrounds, could Dr Oliver tell us whether it is possible to distinguish different backgrounds?

C. J. OLIVER. There are two independent types of information about the background clutter in SAR images: (i) the average cross section and (ii) the texture, described by the surface autocorrelation function. In my presentation I concentrated on the structural (texture) information rather than the cross section. However, both of these properties can be used to distinguish between clutter classes though some texture information will only be available in images from high-resolution systems. One problem is that the mean cross section of fields, for example, may be greater or less than that of the woods. Another is that the texture variation of woods overlaps with typical field textures. Thus it is not practical to classify the different regions unambiguously. However, we can separate images into regions of the same intensity within the statistics (i.e. segmentation) more reliably. It seems probable that the next important stage is to combine SAR images with map-like prior knowledge to address questions such as 'given that this region is a field what information can one deduce about the crop growing in it?' No convincing correlation between images and crop identification has as yet been generally demonstrated. Even more specific prior knowledge may be required such as defining the crop and asking how well it is growing, for example.

J. A. ALLAN. Do these procedures require massive computing power?

C. J. OLIVER. The type of computing power needed for reasonable response times of about a second in the analysis that I have been describing would be provided by, for example, a transputer-based architecture using about 30 transputers delivering about 30 Mips (million instructions per second). Whether this is massive depends on perspective! The methods I outlined were ideally suited to this type of parallel-architecture implementation that is currently advancing dramatically, particularly in terms of unit cost.

Phil. Trans. R. Soc. Lond. A **324**, 409–421 (1988)
Printed in Great Britain

Automatic interpretation strategies for synthetic aperture radar images

By S. Quegan[1], A. J. Rye[2], A. Hendry[2], J. Skingley[2] and C. J. Oddy[2]

[1] *Department of Applied and Computational Mathematics, The University, Sheffield S10 2TN, U.K.*
[2] *GEC Research Ltd, West Hanningfield Road, Gt. Baddow, Chelmsford, Essex CM2 8HN, U.K.*

[Plate 1]

Effective use of the large amounts of timely data to be provided by the coming generation of space-based radars will require automatic methods of image interpretation. The key to such interpretation is an image representation, based on low-level operations, which can support the introduction of high-level (rule-based) knowledge. This representation, described in this paper as a segmented image database (SID), is dependent on the performance of the low-level operations (segmentation, edge-detection and thin-line-detection) which generate it. Methods of quantifying the performance of these operations are described. Use of the SID to support classification based on context, and image–map matching that uses image structure, rather than geometrical matching, are demonstrated.

1. Introduction

The use of synthetic aperture radar (SAR) to investigate and monitor the Earth's surface is of increasing interest, because it can provide images with high spatial resolution unaffected by cloud or darkness, and it responds directly to motion in the observed scene. High resolution (of the order of metres) is important in many civilian and military applications.

Penetration of cloud is essential for routine land-use monitoring in, for example, Europe or the equatorial rain forests. The ability to operate through cloud and at night is particularly relevant at high latitudes, where there is a need to monitor ice and icebergs (Skriver & Gudmandsen 1986). The Doppler shift imposed on the return signal by a moving scatterer can be used to selectively image moving small objects (Raney 1971; Freeman 1984) or to image waves in the ocean (Rotheram 1983).

These attractive possibilities offered by SAR have provided the impetus to several major satellite projects for Earth observation, intended to be implemented before the end of the century. The European Space Agency, Japan and Canada are all developing free-flying SARS (*ERS*-1, *JERS*-1, and *Radarsat* respectively); the Space Shuttle has been used to carry SARS (the *SIR*-A and *SIR*-B missions), and should continue to do so when missions resume; and a SAR is seen as an essential element in the Space Station–Polar Platform.

The unique capabilities of SAR data unfortunately exact a price, whose value we do not fully know; this is learning how to interpret the data, which has special properties inherent in the use of coherent radiation with wavelengths of the order 1–30 cm. Tonal and textural properties of the images reflect volume and surface scattering in the elements of the scene. Image geometry is intimately connected with relative motion of the scene and SAR platform. Lastly, and of particular importance for the work presented below, SAR images are affected by a signal-dependent noise known as speckle (for a discussion of this phenomenon see, for example, Dainty

(1984)). Speckle is the first obstacle to be overcome in any attempt to interpret SAR images automatically.

Image interpretation has four basic elements:

(i) definition of objects of interest, e.g., fields, texture, areas of change, edges;
(ii) characterization of objects of interest *in image terms*;
(iii) use of (ii) to derive methods of seeking objects of interest;
(iv) referencing of the image and objects within it in a form suitable for comparison with other data.

Element (i) has the obvious implication that image interpretation is application-dependent. However, certain operations seem essential to many applications, notably image segmentation, edge-detection, line-detection, and post-segmentation operations, such as measurement of segment statistics and segment classification. Element (ii) can present great difficulties. Even such primitive concepts as lines, which are handled effortlessly by the eye and brain, seem to defy any simple description in terms of image statistics or elementary structures. Measurement of image properties is often not easy, because it is not clear what should be measured. However, there have been some successes; e.g., image texture has been successfully characterized by combining measurements with an image model (Oliver 1986). Element (iii) is often hindered by the difficulty in producing proper image characterization, and we are often obliged to use paradigms to drive algorithm development, with ensuing questions over the applicability of the results to real images.

Element (iv) is fundamental to use of the data. It is unlikely that the image is the only or even the primary data source of interest. Hence it is essential that the image structure be encoded in a form suitable for comparison and combination with other data. This encoding may be in terms of pixels (with ensuing requirement for geometric corrections) or in terms of a relational database. The generation and use of such a database is a central concern of this paper, because it provides the framework for introduction of knowledge-based rules into image interpretation. The low-level operations used to generate the database are discussed in the first half of the paper. The emphasis here is on quantifying and optimizing the performance of algorithms, thus defining the constraints in the data available to the database. The second half of the paper is concerned with the use of the database to compare images, and to use contextual information in classifying image segments.

2. Low-level image processing

By low-level image processing we will mean those operations necessary to generate a segmented image database, i.e. edge detection, thin-line detection and image segmentation. Other potentially important operations, such as statistical correlation algorithms, are not discussed (but see Deane *et al.* 1987). Although describing these operators as low level, we do not wish to imply that they are easy to develop or do not rely on sophisticated techniques. The opposite is in fact the case, and the search for reliable methods operating in speckle still continues. In the next section we summarize our approaches to quantitative evaluation of algorithms.

2.1. *Edge detection and segmentation*

As our working definition of an edge, we use a sharp discontinuity in scene reflectivity, which becomes smoothed as a result of the point spread function of the SAR imaging process

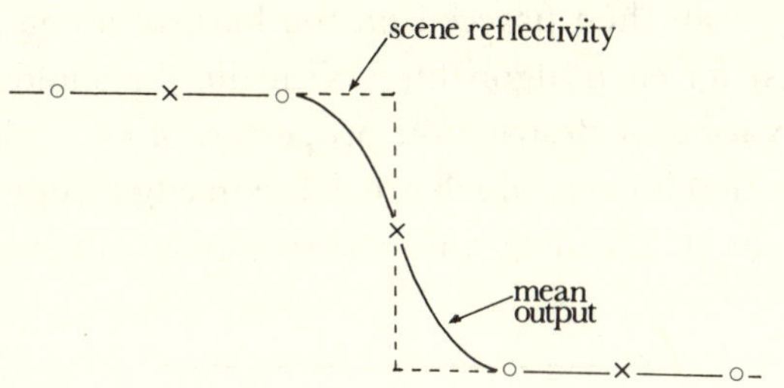

FIGURE 1. Mean output in the SAR image of a sharp reflectivity edge. Two possible digital samplings are noted by ○ and ×. The mean intensity indicated would be subject to speckle.

(figure 1). Edges and segments are dual concepts, each giving rise to the other, and segmentation can be carried out by edge-detection, segment growing or a combination of both approaches (White 1985; Oddy & Rye 1983).

An obvious question when faced with the large variety of methods mooted for image segmentation concerns their relative merits and reliability. Because this is fundamental to operational use of these methods, we have sought to define procedures to compare different segmentation schemes. This has relied extensively on simulated images, because (*a*) the definition of the correct or desirable segmentation of a real image is not available, and (*b*) segmentation procedures normally involve nonlinear and/or adaptive operations which are not easy to investigate analytically. Though there are a number of properties of segmentation that may be regarded as desirable (e.g. area-preservation, edge position), we have concentrated on the two most fundamental properties, i.e. the generation of spurious segments (false alarms) and the merging of two distinct segments (edge-destruction). It has at present only proved possible to produce a method of comparing segmentation algorithms of the form shown in scheme 1.

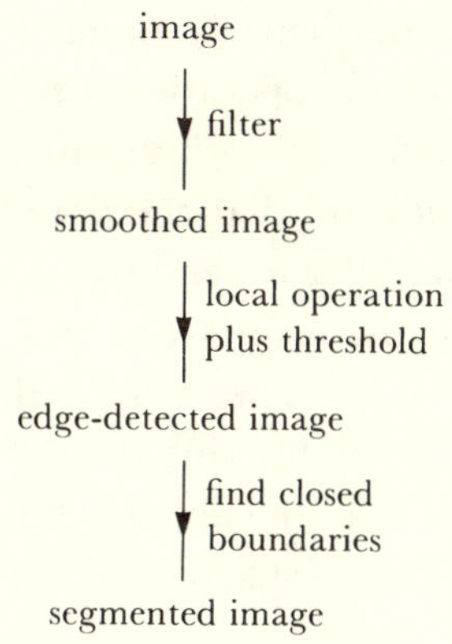

SCHEME 1. Edge-detection segmentation scheme.

This omits the important methods based on segment growing, and such possibly important techniques as iterated filtering before edge detection. At present we do not know how to extend our methods to these cases, except by a massive simulation programme.

The approach used is very simple in concept. Because edge-detection thresholds the output of some operation applied to the filtered image (see scheme 1), a false-alarm measure can be derived by examining the effects of applying the operation to filtered pure speckle images. This yields, for a particular algorithm, a relation between false-alarm rate and the threshold used.

Comparison of algorithms can then proceed on the basis of a comparable false-alarm rate (and hence fixed thresholds) for each algorithm. With the thresholds fixed, it is necessary to investigate the edge preservation or destruction properties of each algorithm. This is carried out by first generating simulated images which are dense in edges (figure 2), and then applying each algorithm to enough simulated images to ensure statistically significant results.

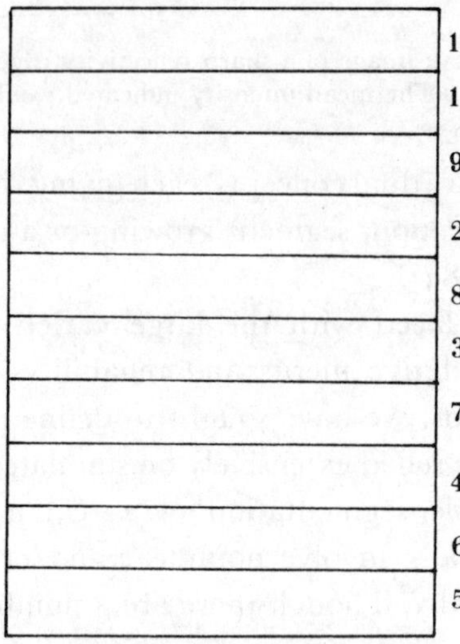

FIGURE 2. Edge image used for simulation. The numbers on the right give the relative brightnesses of the horizontal bands. Speckle is added before simulation is carried out.

These methods have been applied to a number of proposed schemes for edge-detection based segmentations, and have proved particularly useful for comparing schemes that have free parameters in the filtering stage of the operation. Of particular interest are results obtained with the Frost *et al.* (1982) filter combined with a matched edge detector (for a fuller discussion see Quegan *et al.* 1987). These indicate that the optimum performance is achieved (for any false alarm rate) by permitting the filter to tend to an average filter. A similar conclusion has been obtained from methods that use a filter due to Oddy & Rye (1983) and the modified Lee 2-σ filter (Lee 1983) (see Hendry *et al.* 1985), with the implications that (i) nothing is gained by greater sophistication in the filtering, hence simple and fast methods can be used, and (ii) the edge-smoothing inherent in the average filter does not prevent post-filtering edge-detection. These conclusions have been tested on real data, with results that confirm the findings from simulated data.

It must be stressed, however, that we cannot as yet define an optimum edge-detector or segmenter, only put limits on the reliability of one class of such algorithms. These algorithms have also been applied to only a limited class of data. The effects of oversampling the data have been investigated (the major conclusion being that any form of oversampling degrades the performance of the algorithms), but edge-orientation effects and the effects of resampling have not been considered. It must also be noted that the test pattern (figure 2) was designed to keep the results at different edges statistically independent, hence the results may not be applicable as the stripes become narrower. There is therefore a continuing need to develop the methods described above, to extend the range of image data for which they may be considered valid.

There is also a clear need to develop methods of assessing the performance of other classes of algorithms, if we are to provide the best techniques for applications purposes.

2.2. *Thin-line-detection*

Thin lines are important scene elements (roads, hedgerows, field boundaries, etc.), which contribute to human image segmentation. We characterize thin lines as features one to three pixels wide, which may separate regions of the same or different intensities, and which may traverse several different regions. Conceptually they could be regarded as segments, but in practice special-purpose algorithms are necessary to locate them, because their statistical separability from the background requires searching along particular directions (both because of their narrowness, and because they often do not differ greatly in intensity from the background in real SAR images).

Three types of line-detection methods have been considered, namely (1) local operators or masks (Vanderbrug 1976; Gurney 1980); (2) dynamic programming (Wood 1985); and (3) global transforms (Deans 1981; Duda & Hart 1972). Of these, the first method seems inapplicable to SAR images unless very large window sizes are used, and it does not perform well when the ratio of line to background intensity is near one (Hendry *et al.* 1985). The second method is applicable to SAR images, can cope readily with curved lines, is computationally expensive and needs further development to cope with linking of line segments (Wood 1985; Skingley 1986). The third method uses the Radon or Hough transforms, and has been successfully applied to SAR images (Murphy 1987; Hendry *et al.* 1985; Skingley 1986). The application of these two transforms to line-detection yields equivalent results (the Hough transform is a special case of the Radon transform). The Radon transform is faster (Murphy 1985), but the Hough transform is conceptually simpler, and can be readily analysed (Skingley & Rye 1987; Skingley 1986), as described below.

The Hough transform parametrizes all possible lines in the image in terms of R, the distance from the centre of the image, and θ, the orientation (see figure 3), and sums pixel values along

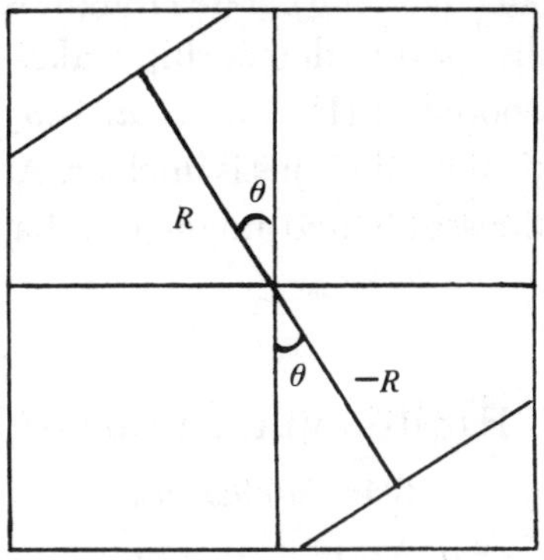

FIGURE 3. The Hough transform coordinate scheme.

these lines, assigning mean intensity of the line to each (R, θ) point. Bright and dark lines in the image therefore transform to peaks and troughs in the (R, θ) domain. Detection probabilities for such peaks and troughs can be derived analytically (see Skingley & Rye 1987). Figure 4 shows calculated detection probabilities for lines of various lengths and brightnesses. Comparison

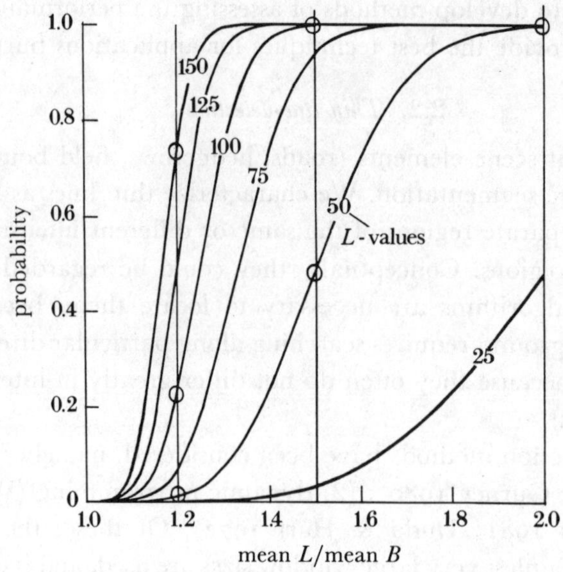

FIGURE 4. Calculated detection probabilities for lines of various lengths (indicated on each curve) embedded in a uniform background, with Rayleigh distributed speckle. The abscissa gives the ratio of line, L, to background brightness, B. The maximum possible line length is taken to be 150 pixels.

with performance on simulated images confirmed these theoretical results, though an excessive number of spurious detections were noted. These can be understood in terms of correlation between samples in the transform space (Skingley 1986).

Detection alone is not enough, because it is clear from the formulation of the Hough transform that short bright lines yield the same signature as long, less bright lines. Work on end-point detectors and removal of false alarms is still continuing.

The applicability of results based on simulation to real data must be carefully considered, because at least some lines observed in real images do not conform to the simple speckle model adopted in the simulation (Quegan *et al.* 1988). The Hough–Radon transform has been applied to real SAR images, and successfully used to detect ship wakes and quite subtle patterns of lines corresponding to rides through woodland (Hendry *et al.* 1985; Murphy 1987). Whether these results are consistent with theoretical predictions is unclear. At present, also, thin-line-detection is not properly incorporated in the segmented image database, and more work is required to accomplish this.

3. High-level processing

3.1. *Introduction*

Segmentation techniques, such as those described above, generate an initial spatial description of a scene in a form suitable for further analysis of the image data. The development of techniques that manipulate and process image regions has been supported by the construction of a standard database of segmented images. This database is described in the next section.

The spatial representation of a scene may be matched to other geographic data, such as maps, to generate a mapping function between the data, as a precursor to the fusion of multisensor and multitemporal images. One approach to matching segmented images is described below.

It has been recognized, for many years, that contextual information is a valuable tool for producing land-use classification from remote-sensing images. However, until recently many classification methods used only the spectral response of single pixels and their neighbours. Segmentation provides the necessary primitives for the development of classifiers that use the contextual information present in a scene and a simple demonstration system for woodland is presented.

3.2. *A segmented image database* (SID)

A database for storing region information from an image after it has been segmented has been designed by GEC Research (Cruse *et al.* 1984). This Segmented Image Database (SID) stores information on regions, region boundaries, adjacencies and attributes in a flexible manner, thus allowing the database to be accessed in a number of ways for different analysis operations. The SID can be easily updated after these operations, and can store images from any source.

Two implementations of the SID have been designed in FORTRAN and PROLOG. The FORTRAN implementation is suitable for low-level image-analysis operations and the PROLOG implementation is suitable for high-level tasks.

In FORTRAN, the database is basically a tree structure, with additional pointers between branches, and uses a dynamic memory management system. The data are contained in three main indexes: a region index, containing all the regions in the image, a line segment index, containing all the line segments that make up the boundaries of the regions, and an attribute index, containing information about all attributes that have been calculated for any of the regions. These attributes may include mean grey level, texture measures, shape measures, polygonal approximations, symbolic labels and any string or vector the user wishes to associate with a region. A suite of subroutines has been written to interact with the SID to create, read, delete or update records.

The network structure of the FORTRAN implementation may be converted to a relational database in PROLOG. This allows very simple facilities for interrogating the database and rapid assessment of new algorithms.

Many different image analysis operations may be performed with the SID, e.g. attribute extraction, region merging, region classification, and shape recognition. It is also easy to create graphical descriptions of the segmented image from SID, such as a region adjacency graph, which can be used for structural matching techniques.

3.3. *Structural matching*

Structural matching is a technique which matches images or images and maps after they have been segmented. The segmented images are represented by attributed region adjacency graphs (RAGS), in which the nodes of the graph represent the regions, and the branches represent the adjacency relations. Both the nodes and the branches of the graph may have attributes associated with them. These may be the size, shape or texture of regions and the direction of adjacency. The images are then matched by matching their RAGS.

This region-based approach to matching will be less sensitive to speckle than methods relying on local correlation, as long as the RAG is of 'sufficient' accuracy. It is not yet possible to relate the performance measures discussed in §2 to a proper description of the accuracy of the RAG. However, the method of matching described below seems capable of coping with highly distorted RAGS, as will be clear. This method may also be used for multisensor and multitemporal image matching when pixel-by-pixel comparisons are not sufficiently robust.

The matching strategy developed by GEC Research (Oddy 1986; Oddy *et al.* 1988) involves first matching a number of control regions from the images. These should be distinctive in some way (for example in size) and ideally should be well distributed throughout the images. The second stage then attempts to match regions adjacent to those already matched, until no more regions can be matched. Thus the matching proceeds outwards from the control regions.

The rules for matching two regions are based on a comparison of region attributes and region shape. Region shape is determined by region vectors from a polygonal approximation of the boundaries.

The matching process allows for partial matching of regions. This is necessary because a complete match between two RAGS is very unlikely, due to genuine differences on the ground between images taken at different times, to different sensors discriminating between different surface types and to imperfect segmentations.

Output from the matching scheme is initially a list of matching region pairs. These can then be used to generate automatically lists of control points which may be used to determine a mapping function from one image to the other image or map. These control points may be the midpoints of matching vectors, or the corners of matching fields.

The above scheme has been implemented in a mixture of FORTRAN and PROLOG on a VAX 11/750. The programs used to generate the SID files containing the attributed RAG are written in FORTRAN. These files are then converted to a PROLOG relational database which is used as input to the structural matching. The matching algorithm is written in PROLOG and runs in the POPLOG system.

The technique described above has been used to match an airborne SAR-580 image (figure 5, plate 1) to a digital map (not shown) of an area near Feltwell in Norfolk. The map was digitized at GEC Research from an ordnance survey 1:25000 map. The map area was originally surveyed in the 1920s, and the SAR image was acquired in 1982; several differences are evident between the two datasets.

The map and the image were segmented using techniques developed at GEC Research (Keen 1986; Cruse *et al.* 1986). Figures 6 and 7 show line drawings of these segmentations.

Matching was performed with the map as the reference image. Rough orientation and scale information were input to the process. As many control regions as possible were then matched, followed by matching of adjacent regions, with gradual relaxation of the shape matching thresholds. The results of this matching are shown in figures 6 and 7. Matching regions are assigned the same letter. In spite of the fact that only a partial segmentation could be generated the matching performed well for regions which were extracted. In particular, pairs A, V and F match well. A number of partial matches are also obtained, in areas that can easily be matched manually (e.g. pairs y, X, N and T) and in more complicated areas (e.g. pairs w, n, u and r).

A set of control points from the midpoints of matching vectors have been generated and used to define an affine transformation from image to map. The outcome was in exact coincidence with the transformation obtained by manual selection of control points, an excellent result in view of the poor segmentation of the SAR-580 image that was provided for the matching process.

FIGURE 5. SAR-580 image near Feltwell, Norfolk.

FIGURE 8. SAR-580 image of woodland area.

(*Facing p.* 416)

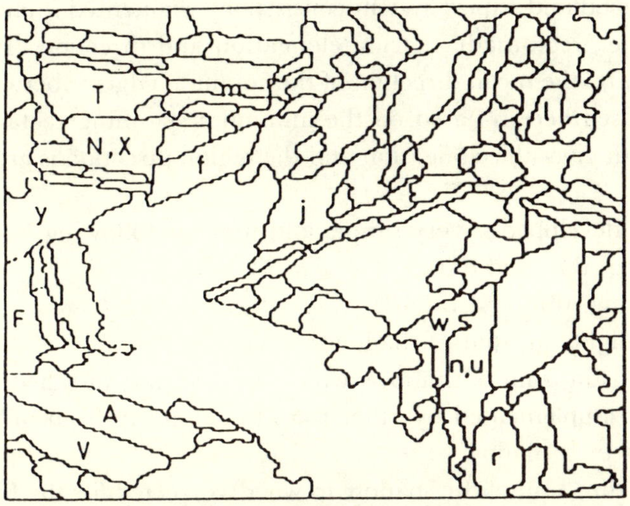

FIGURE 6. Segmentation of lower left section of figure 5 (the sharp river bend is in the top right of the segmentation).

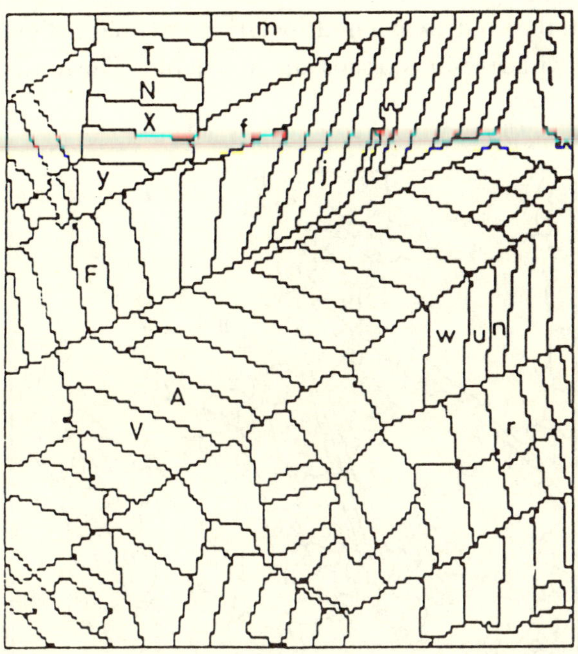

FIGURE 7. Segmentation of the ordnance survey digital map corresponding to figure 6.

3.4. *A simple contextual expert system for woodland*

This section describes a simple demonstration expert system which uses contextual and structural information to identify hedges and woodland in a high-resolution SAR-580 image.

High-resolution (3 m) airborne SAR images often contain shadow and highlight regions when imaging woodland and hedgerows. These linear features are typically 2–3 pixels wide and are often partly detected by segmentation algorithms as elongated regions. A system that uses a number of simple rules to interpret a scene has been implemented in PROLOG by using POPLOG on a VAX11/750.

The input to the woodland expert system consists of a segmented image which includes the region mean grey level, region adjacencies, elongation and direction. The region adjacency directions are defined by the mean direction of the common edge(s) between two regions. The elongation measure used here is based on the minimum spanning rectangle of a region; the ratio of length to width gives the elongation and the region direction is given by the major axis of the rectangle.

The expert system developed is very simple and uses the following basic rules:

1. shadows are long, thin and dark;
2. highlights are long, thin and bright;
3. hedges are highlights in front of shadows;
4. woods are regions in front of shadows which are not part of a hedge.

These basic rules are supplemented by other rules to define 'in front of' for example. Thus:

a wood is IN FRONT OF a shadow

IF the adjacency direction of the shadow to wood (corrected for the flight direction) is less than 75°

AND the azimuth length of the shadow overlaps at least 30% of the azimuth length of the wood.

The test image used in this study was a multipolarization image (figure 8, plate 1) of a rough textured woodland area with tree-lined roads and heathland. The initial segmentation is shown in figure 9.

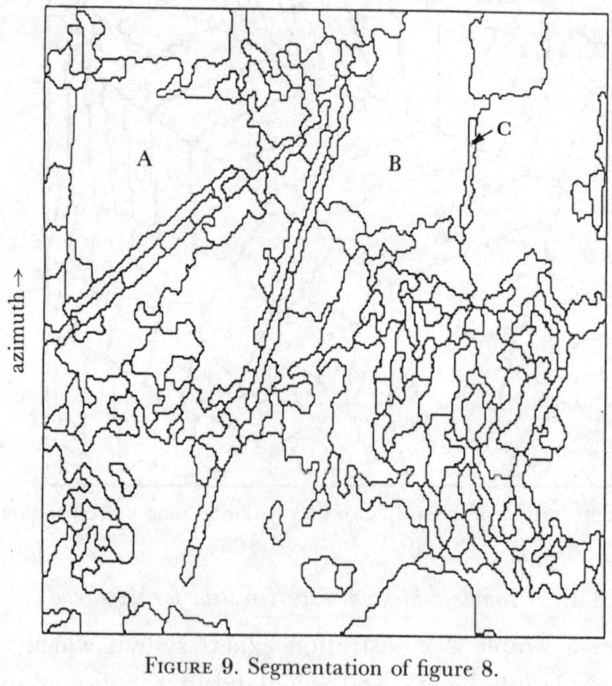

FIGURE 9. Segmentation of figure 8.

First results from the system are shown in figure 10. The hedges are clearly delineated. It is interesting to note that one road is imaged as a shadow in front of a highlight whereas the other road is a highlight in front of a shadow. The first shadow implies a wood in region A but a map of the area suggests that this is not correct. Region B is shown as a wood, because shadow C

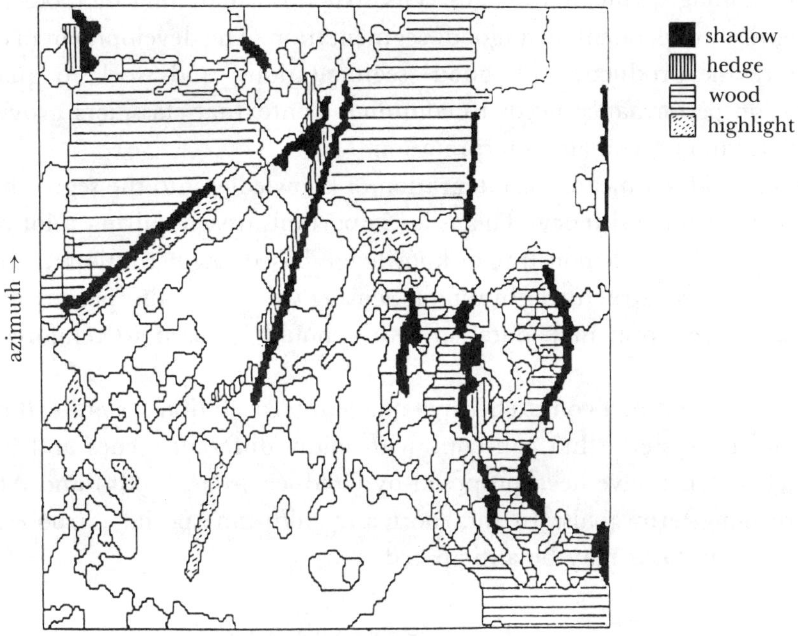

FIGURE 10. Classification of figure 8 using context.

is detected by the segmentation, but the highlight or hedge next to it in the image is missed by the segmentation and region B is incorrectly classified. The interpretation of the textured woodland area, in centre right of the image, is promising.

This woodland expert system shows potential, is very simple and demonstrates how even a small number of rules can be used to assist the interpretation of an image. The interpretation given here relies totally on the structural information in a segmented image and its performance is dependent on the quality of the initial segmentation. Most of the failures of the system are due to a failure by the segmentation scheme to detect some very weak features (highlights and shadows) in the image. In some cases it is very difficult to interpret the original image manually without prior knowledge of the test site or a map of the area.

4. Conclusions

Automatic interpretation of SAR images requires an image representation to support the introduction of knowledge into the interpretation. The segmented image database (SID) provides such a representation. Its formation relies on low-level operations, whose performance imposes constraints on the inferences possible from the SID. Quantitative performance measures for some of the necessary low-level operations are possible. In the case of segmentation based on edge detection, these measures indicate how to optimize the behaviour of the algorithms considered. The methods adopted do not apply to all segmentation strategies, and further work is required. Similarly, the methods allowing a theoretical analysis of the Hough transform as a line-detector cannot obviously be applied to other detection algorithms.

The use of the SID as a means of introducing knowledge into the interpretation of SAR images is at a very early stage. However, the work that has been carried out demonstrates the potential offered by this technology. It has been shown that the matching of image and maps may be achieved by using segmented images. The structural matching methods appear to be robust and not entirely dependent on a 'good' segmentation. The development of contextual classifiers *does* require the production of 'good' segmentations, and work to quantify and improve segmentation performance needs to continue. Contextual classifiers provide important additional information for image interpretation.

The SID provides a means for integration of knowledge into the segmentation process, i.e. it can check its own consistency. This is an important area requiring a lot more work. Nazif & Levine (1984) have demonstrated knowledge-based segmentation methods using general knowledge of object structures. In remote sensing we can obtain specific, if noisy, knowledge of ground structures from map data, and this should be used to reinforce and support edge detection.

The development of a comprehensive image interpretation system will require a number of different expert systems that can cope with many different scenes and image interpretation tasks. Such systems have been proposed by Goldberg *et al.* (1985) and ARTS-IP (1986) but are clearly long-term goals. In the short and midterm special purpose systems for specific applications and tasks may be anticipated.

References

ARTS-IP 1986 Adaptive real time strategies for image processing. Report on Esprit Project 867.
Cruse, D., Oddy, C. J. & Wright, A. 1984 In *Proc. 7th ICPR*, pp. 494–496. IEEE Computer Society Press.
Cruse, D., Hendry, A., Rye, A. J. & Barnes, I. 1986 Marconi Tech. Rpt. MTR 86/62.
Dainty, J. C. (ed.) 1984 *Laser speckle and related phenomena*. Berlin: Springer-Verlag.
Deane, G. C., Quegan, S., Churchill, P. N., Hendry, A., Evans, R. & Trevett, J. W. 1987 In *Proc. ESA Workshop on SAR Applications, Frascati*, SP-264, pp. 95–102.
Deans, S. R. 1981 *IEEE Trans.* **PAMI-3**, 185–188.
Duda, R. O. & Hart, P. E. 1972 *Communs Ass. comput. Mach.* **15**, 11–15.
Frost, V. S., Stiles, J. A., Shanmugan, K. S. & Holtzmann, J. C. 1982 *IEE Trans.* **PAMI-4**, 157–165
Freeman, A. 1984 Proc. IGARSS-84, ESA Pubn. SP-215, pp. 65–70.
Goldberg, M., Goodenough, D. G., Alvo, M. & Karan, G. M. 1985 *Proc. IEEE* **73**, 1054–1063.
Gurney, C. M. 1980 *IEE Trans.* **GE-18**, 204–211.
Hendry, A., Quegan, S., Rye, A. J., Skingley, J. & Churchill, P. 1985 Marconi Tech. Rpt MTR 84/59.
Keen, A. 1986 GEC Technical Note PALS/WN 86.
Lee, J. S. 1983 *IEEE Trans.* **SMC-12**, 83–89.
Murphy, L. M. 1987 *Patt. Recog. Lett.* **4**, 279–284.
Nazif, A. M. & Levine, M. D. 1984 *IEEE* **PAMI-6**, 555–577
Oddy, C. J. 1986 Post segmentation image–image matching. *Marconi Tech. Memo* ITM 85/119.
Oddy, C. J. & Rye, A. J. 1983 *Patt. Recog. Lett.* **1**, 443–449.
Oddy, C. J., Skingley, J. & Howard, P. D. 1988 In *Proc. ESA workshop on Image rectification techniques for space-borne SAR*. (In the press.)
Oliver, C. J. 1986 *Inverse Problems* **2**, 481–518.
Quegan, S., Lamont, J. M., Miller, R. J., Meadows, P. J., Wright, A., Veck, N. J., Churchill, P. N., Holmes, G., Trevett, J. W., Sloggett, D. & Hamer, P. 1987 In *Proc. ESA Workshop on SAR Applications, Frascati*, SP-264, pp. 37–45.
Quegan, S., Hendry, A. & Skingley, J. 1988 In *Proc. IMA Conf. Mathematics and its Applications in Remote Sensing*. (In the press.)
Raney, R. K. 1971 *IEE Trans.* **AES-7**, 3, 499–505.
Rotheram, S. 1983 In *Satellite microwave remote sensing* (ed. T. Allen), ch. 10. New York: Wiley.
Skingley, J. 1986 Marconi Tech. Rpt. MTR 86/72.
Skingley, J. & Rye, A. J. 1987 *Patt. Recog. Lett.* **6**, 61–67.
Skriver, H. & Gudmandsen, P. 1986 Final Rpt to ESA, Contract No. 5442/83/D/IM(SC).
Vanderbrug, G. J. 1976 *IEEE Trans.* **GE-14**, 37–44.
White, R. G. 1985 RSRE Memo no. 3900.
Wood, J. W. 1985 RSRE Memo no. 3841.

Discussion

D. Lane (*Intelligent Automation Laboratory, Department of Electrical and Electronic Engineering, Heriot-Watt University, Edinburgh, U.K.*). The effectiveness of a segmentation strategy usually depends upon the appropriate selection of one or a number of parameters, such as thresholds, for a given class of data. A truly automatic interpretation strategy should not rely on any manual input of these parameters. What prospect is there, therefore, for automatically selecting processing parameters, by using perhaps high-level knowledge?

S. Quegan. The first part of the paper described an approach to defining optimal parameters on purely statistical grounds, making use only of general properties of the data. For an initial segmentation such optimized procedures seem appropriate; they should not appeal to high-level knowledge, but to known properties of the data, based on the features of the sensor. The methods described offer hope that optimal parameters can be defined, though at present we can only cope with a limited class of algorithms and of data. Where high-level knowledge should surely be introduced is in assessing and perhaps adjusting the initial segmentation. This may involve reprocessing with altered parameters, but may instead involve closer scrutiny of selected areas, or imposition of rules to refine the segmentation. That is an interesting and difficult area of study.

Phil. Trans. R. Soc. Lond. A **324**, 423–435 (1988)
Printed in Great Britain

Building knowledge-based systems for detecting man-made structures from remotely sensed imagery

By D. M. McKeown, Jr

Department of Computer Science, Carnegie-Mellon University, Pittsburgh, Pennsylvania 15213, *U.S.A.*

[Plates 1–3]

In this paper we review and discuss several emerging themes in the area of image-interpretation for automated cartography and digital mapping. Our primary observations are that the detection, interpretation, and analysis of man-made structures by remotely sensed imagery requires the integration of spatial knowledge with image-analysis and interpretation techniques. General knowledge of structural and spatial layouts for man-made structures, as well as *a priori* map knowledge, can be expected to constrain search during image analysis. Spatial databases consisting of geographically referenced facts should form the basis of *a priori* knowledge necessary to perform interpretation and have a dual role as the repository of partially compiled facts acquired during scene-analysis.

1. Introduction

Knowledge-based interpretation of remotely sensed data requires knowledge about the scene under consideration. Knowledge about the type of scene (airport, suburban housing development, urban city) can aid in low-level and intermediate-level image-analysis, and can be expected to drive high-level interpretation by constraining search for plausible consistent scene models. Loosely speaking, this knowledge base should contain known facts and spatial relations between objects in an area of interest, access to historical or *a priori* map knowledge, and methods to relate Earth coordinates to pixel locations in digital imagery.

In this paper we describe three experiments in knowledge-based interpretation. Each experiment uses spatial and structural knowledge at an increasingly higher level of detail. These experiments can be characterized as narrowly defined, vertically integrated image-analysis systems used to extract specific features such as roads, buildings and complex sites such as airports and urban environments. These programs are not 'general vision' systems because they capitalize on the task-specific nature of the problem domain to constrain the complexity of the world. Nevertheless, they are important datapoints in a domain where many important questions concerning knowledge representation and use largely remain unanswered.

The primary role for knowledge-based systems in the analysis of remotely sensed imagery is to provide constraints so that image analysis and interpretation tools can be used in spite of their inherently error-prone performance. The goal then, is to integrate rule-based systems with image-analysis techniques to constrain the search space of possible scene interpretations by using domain and general knowledge. These constraints can be characterized as 'what to look for' and 'where to look for it'. Currently, knowledge-based systems are most powerful (and successful) in narrow, well-defined task areas. It is certainly the case that to solve the general remote-sensing problem an analysis system will require use of general problem-solving capabilities and vast amounts of domain and common-sense knowledge, currently far beyond the capability of any research system. However, even without general problem-solving capabilities, there is much to be gained by the development of 'existence proof' knowledge-based interpretation systems. Current image-processing systems are incapable of high-level description of the results of image interpretation. The combination of a map-based world model and knowledge-based systems which have site-specific or task-specific knowledge can be used to bridge the gap between users and current state-of-the-art image-interpretation systems. The long-term goal of our research is to develop systems that can interact with a human cartographer or analyst at a highly symbolic level, maintain a database of previous events, and use expert-level knowledge to predict areas for fruitful analysis, and integrate the results of the analysis into a coherent model.

In the remainder of this paper we describe three experimental analysis systems that use low-level, intermediate-level and high-level spatial knowledge, respectively, to extract and interpret man-made structures from aerial imagery. These systems are MACHINESEG, a system for map-guided image segmentation, ARF, a system for finding and tracking roads, and SPAM, a rule-based system for airport scene interpretation. Each is linked by the common use of a spatial database system, MAPS/CONCEPTMAP, to provide access to and information concerning image, map, and terrain data. In the following section we briefly describe the MAPS database representation for spatial knowledge.

2. An overview of MAPS

The MAPS spatial database (McKeown 1983, 1984*a*, *b*, 1987) was developed between 1980 and 1984, supported by the DARPA Image Understanding Program as research into large-scale spatial databases and spatial knowledge representation. It is interesting that this system has expanded from its original research goal of developing an interactive database for answering spatial queries into a component of several knowledge-based image-understanding systems under development at Carnegie-Mellon University. MAPS is a large-scale image–map database system for the Washington, D.C. area that contains approximately 200 high-resolution aerial images, a digital terrain database and a variety of map databases from the United States Geological Survey (USGS) and the Defense Mapping Agency (DMA). MAPS has been used as a component for an automated road finder–follower, a stereo verification module, a 3-D scene-generation system, and a knowledge-based system for interpreting airport scenes in aerial images. In addition, MAPS has an interactive user-query component that allows users to perform spatial queries with high-resolution display of aerial imagery as a method for indexing into the spatial database. This capability to relate, over time, imagery at a variety of spatial resolutions to a spatial database forms a basis for a large variety of interpretation and analysis tasks such as change detection, map update and model-based interpretation.

Figure 1 shows the system organization of MAPS. Four databases are maintained within MAPS: a digital terrain database, a map database, a landmark database and an image database. A fifth database, CONCEPTMAP, consists of a schema-based representation for spatial entities and a set of procedural methods that provide a uniform interface to each of the four component databases for interactive users or application programs. It is this interface that allows us to represent and access image, map, terrain and collateral data in a manner that best suits the intrinsic structure of the data. At the same time the CONCEPTMAP databases provides uniform access to a variety of spatial data independent of the particular internal structure. This is in

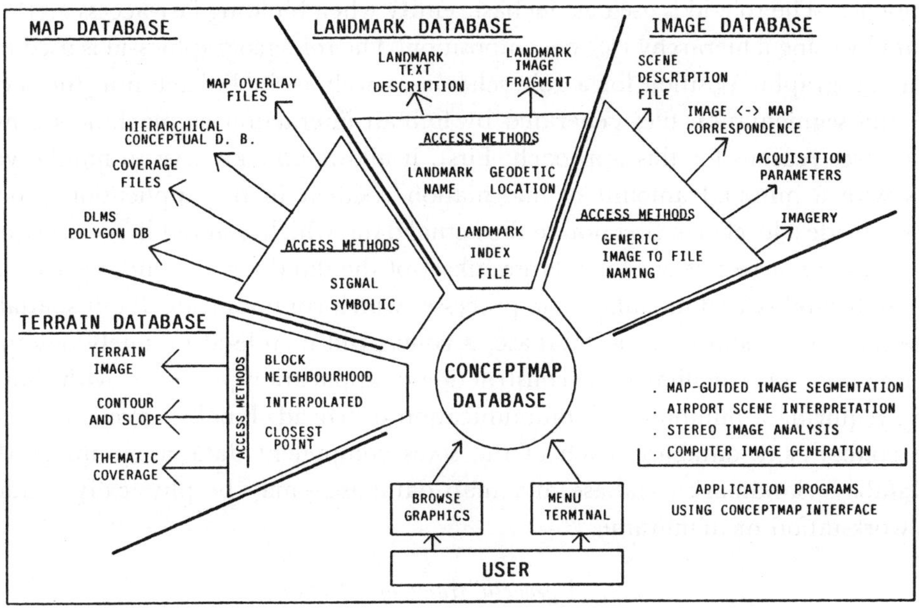

FIGURE 1. MAPS: system overview.

sharp contrast to methods proposed for uniform representation of image and cultural data such as raster data sets and regular decompositions such as quadtrees or k–d trees.

2.1. *Schema-based representation for spatial entities*

The CONCEPTMAP database uses a schema-based representation for spatial entities. The use of schemas (or frames) is a well-understood AI methodology for representing knowledge. Such a representation can be combined within several problem-solving methods such as semantic networks, scripts or production systems to construct a problem-solving system (Barr & Feigenbaum 1981). Each entity in the CONCEPTMAP database is represented by one concept schema and at least one role schema. A concept can represent any spatial object and associates a name with a set of attributes stored in the concept and role schemata. A concept such as 'washington d.c.' might have multiple role schemata defined, each representing a different view of the same spatial area, such as 'political', and 'demographic'. A concept such as 'georgetown university' might have multiple role schemata defined for each of the campus buildings or areas.

There are three unique identifiers generated by the CONCEPTMAP system which allow for indirect access to additional factual properties of concept or role schemata.

(i) The concept-id is unique across all concepts in all CONCEPTMAP databases. That is, given a concept-id one can uniquely determine the name of the spatial entity.

(ii) The role-id uniquely determines a role schema across all CONCEPTMAP databases.

(iii) The role-geographics-id uniquely determines a collection of points, lines or polygons in vector notation. Each point is represented as $<$ latitude, longitude, elevation $>$.

These identifiers are also used to index into other components of the MAPS database. For example, the concept-id is used to search for landmark descriptions of measured ground control points used during the calculation of transform functions for image-to-map and map-to-image correspondence. The role-id is used as the basic entity when building fast-access methods to the spatial data by using a hierarchy tree decomposition. The role-geographics-id is used to acquire the unique geographic position for a role schema as well as for linkage into the MAPS image database and segmentation files generated by human interaction or machine segmentation. There are three reasons for this approach. First, it allows CONCEPTMAP to handle very large databases with a minimal amount of information resident in the application process. The identifiers provide a level of indirection to the actual data, which is stored in a variety of formats and may or may not be present for a large subset of the database. Second, we can achieve a great deal of flexibility and modularity in processes which communicate about spatial entities. Given the name of a CONCEPTMAP database, a concept-id or role-id uniquely determines the entity in question. This facilitates the construction of application programs with simple query structures, requiring a minimum of communication overhead. Finally, given this decoupling from the CONCEPTMAP database, each of the MAPS component databases – image database, terrain database, landmark database and map database – may be physically resident on a different workstation or mainframe.

2.2. *A geodetic frame of reference*

An implicit requirement crucial to successful application of spatial knowledge for image analysis is that the metrics used by the analysis system be defined in cartographic coordinates, such as < latitude, longitude, elevation >, rather than in an image-based coordinate system. Systems that rely on descriptions such as 'the runway has area 12000 pixels' or 'suburban homes are between 212 and 345 pixels in area' are useless except for (perhaps) the analysis of one image. Further, spatial analysis based on the semantics of above, below, left-of, right-of, etc., are also inappropriate for general interpretation systems. To apply metric knowledge, one must relate the world model to the image under analysis. This should be done through map-to-image correspondence with camera models. We can directly measure ground distances, areas and absolute compass direction, and recover crude estimates of height using a camera model computed for each image under analysis. The MAPS system allows us to express spatial knowledge in ground-bases metrics, which in turn allows us to work with imagery and map data at a variety of scales and resolutions.

In the following sections we describe the first of our three experimental analysis systems, MACHINESEG. It can be characterized as using low-level spatial knowledge, primarily two-dimensional shape information, coupled with general positional information to guide region analysis and segmentation.

3. Map-guided image segmentation

MACHINESEG (McKeown & Denlinger 1984) is a program that performs map-guided image segmentation. It uses map knowledge to control and guide the extraction of man-made and natural features from aerial imagery using a region-growing image-analysis technique. MACHINESEG uses the CONCEPTMAP database from the MAPS system as its source of map knowledge. Figure 2 shows the interaction between the map database and image-processing and feature-extraction tools. Map knowledge can be used to represent generic shapes, sizes and spectral properties typical to a large class of objects such as roads, or can describe specific features such as known buildings, where geodetic position may be known as well as the more

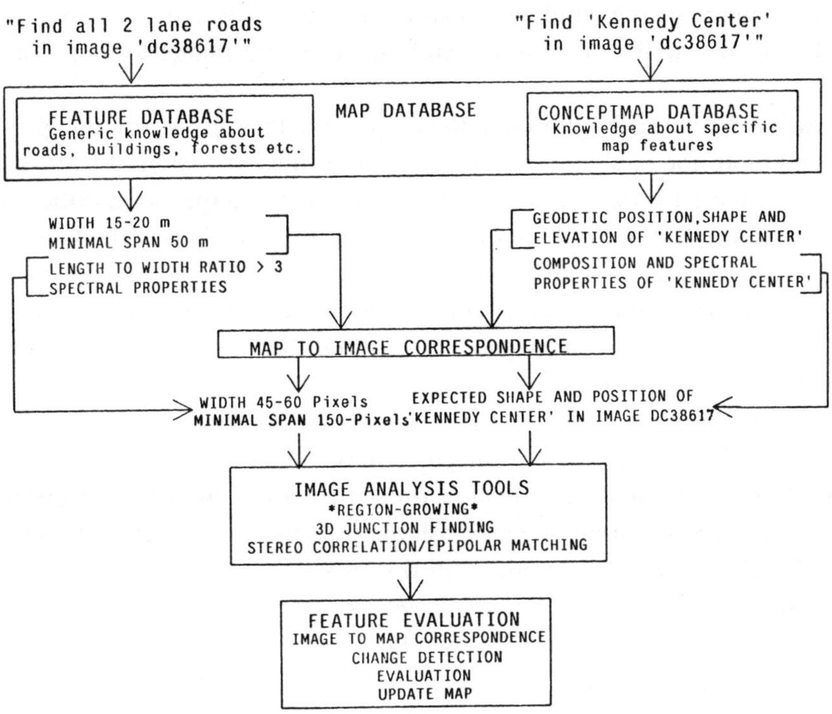

FIGURE 2. MACHINESEG: map-guided feature extraction.

general structural properties. For the latter we have been able to segment a diverse set of cultural features such as buildings, reservoirs and roads, using map-to-image correspondence to project map-based descriptions on to a new image under analysis. This projection generally only provides a coarse idea of the actual position of the feature, but greatly constrains search in the image. MACHINESEG uses the following components of the MAPS system.

(i) CONCEPTMAP to retrieve shape and position models.

(ii) Map-to-image correspondence to project models on to new imagery and image-to-map correspondence to calculate metric distances and areas.

(iii) CONCEPTMAP to store extracted features from several images before interactive editing and integration into the database.

It is important to characterize what we mean by 'map-guided' image segmentation. Map-guided image segmentation is the application of task-independent spatial knowledge to the analysis of a particular image, using an explicit map-to-image correspondence derived from camera and terrain models. Map-guided segmentation is not interactive editing or computation of descriptions in the image domain, because these descriptions are valid for only one specific image. As we have described, each role schema contains a role-geographic-id which points to a geodetic description (< latitude, longitude, elevation >) for each map entity in the CONCEPTMAP database. This description is in terms of points, lines and polygons, or collections of these primitives. Features such as buildings, bridges and roads have additional attributes describing their elevation above the local terrain, as well as their composition and appearance. The location of each map feature in the database can be projected on to a new image by a map-to-image correspondence maintained by MAPS. Likewise, a new map feature can be projected on to the existing image database. If camera-model errors are known, one can directly calculate an uncertainty for image-search windows. Further, as new features are acquired their positions can be directly integrated into the map database by using image-to-map correspondence procedures.

Figure 2 shows a schematic description of the map-guided feature-extraction process in MAPS. There are two methods for applying map knowledge to the extraction of features from aerial imagery. The first method uses generic knowledge about the shape, composition and spectral properties of man-made and natural features. This may be provided by the knowledge base of the application. The second uses map-based template descriptions. These descriptions are stored in the CONCEPTMAP database and represent knowledge about known buildings, roads, bridges, etc. This knowledge includes geodetic position, shape, elevation, composition and spectral properties. In the second case, the position, orientation and scale are constrained, whereas in the first, only the scale can be determined. For both to operationalize spatial knowledge for the analysis of a particular image, a map-to-image correspondence is performed.

Photographs shown in figure 3, plate 1, show the sequence of region-merging steps performed automatically by MACHINESEG using a generic shape description for a set of buildings near the Ellipse park in Washington, D.C. Photograph 0 shows the original image. Photographs 1–5 show how the region merges are evaluated against the shape description, those regions satisfying the shape criteria being marked as solid filled polygons. While a large number of potential regions are generated from the initial image (1), only a small number correspond to buildings. Photograph 5 shows the final result of all seven buildings correctly segmented, with two incorrect segmentation regions generated.

It should be stated that MACHINESEG does not have a model of buildings, or the relation of buildings to roads, or any particular high-level description of the image under analysis. Such models or expectations must be provided by other processes, such as an interactive user, or an analysis system. What makes MACHINESEG a powerful image-analysis tool is that it can use shape descriptions and search among the region segmentation for those regions that best match the segmentation in terms of shape, size and orientation properties. Thus, MACHINESEG is a good example of a low-level generator of plausible regions that must be analysed by intermediate and high-level processes within the context of a particular task. It can be argued that more specific knowledge about the structure of buildings and roads should be brought to bear at the low level. Our philosophy has been to avoid that feature-specific approach during the segmentation phase and to defer such analysis until a coarse overview of the entire scene can be accomplished.

In the following section we describe a system, ARF, that produces intermediate-level and high-level descriptions of roads in high-resolution aerial imagery. It uses knowledge about the structure of roads and employs multiple methods to improve overall system performance. In contrast to MACHINESEG, which can also be used to segment linear features, such as roads, railroad lines and certain drainage patterns, ARF relies on a very specific model of roads to achieve a much higher reliability and performance levels than could be achieved with a general region-based system such as MACHINESEG.

4. Cooperative methods for road tracking

Road finding in remotely sensed imagery has often been equated with linear feature extraction. The rationale was that finding linear features in imagery either by region extraction or line finding was equivalent to finding roads. Further, line finding also worked for other types

FIGURE 3. Sequence of image segmentation for small buildings.

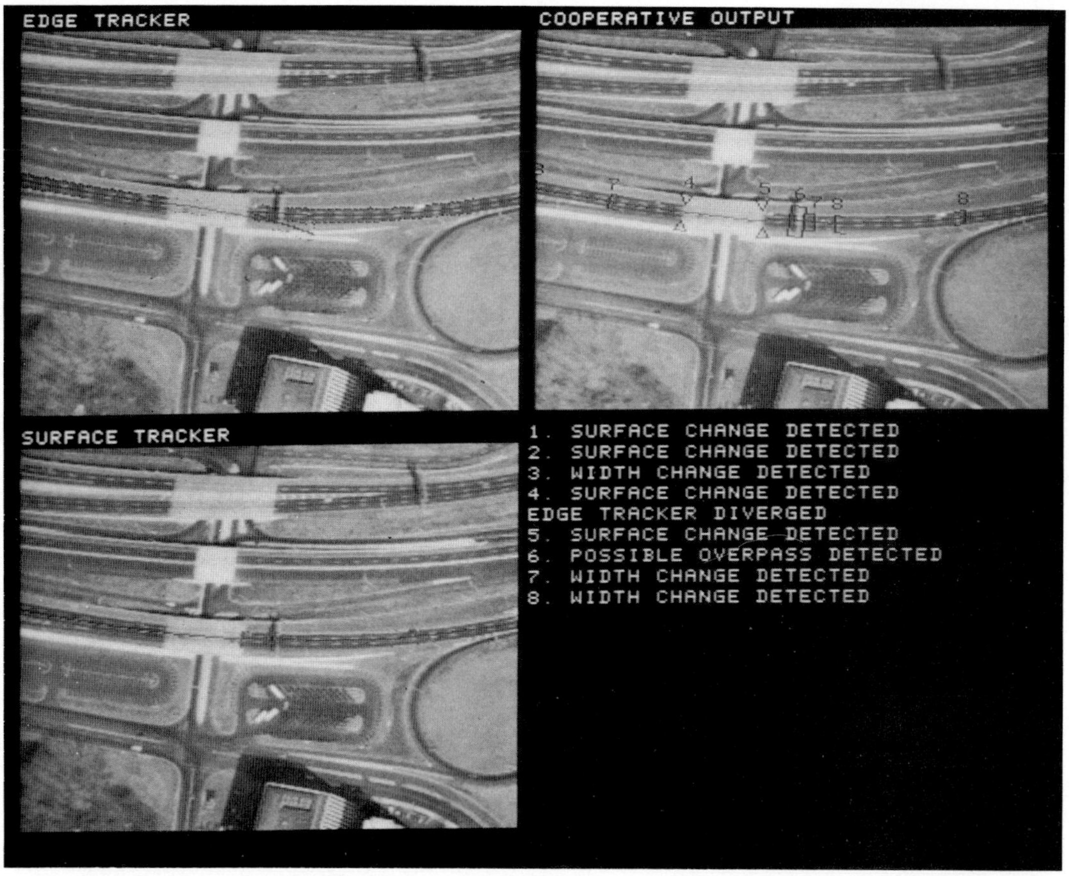

FIGURE 5. ARF: edge tracker fails on low-contrast overpass, successfully restarted from surface tracker.

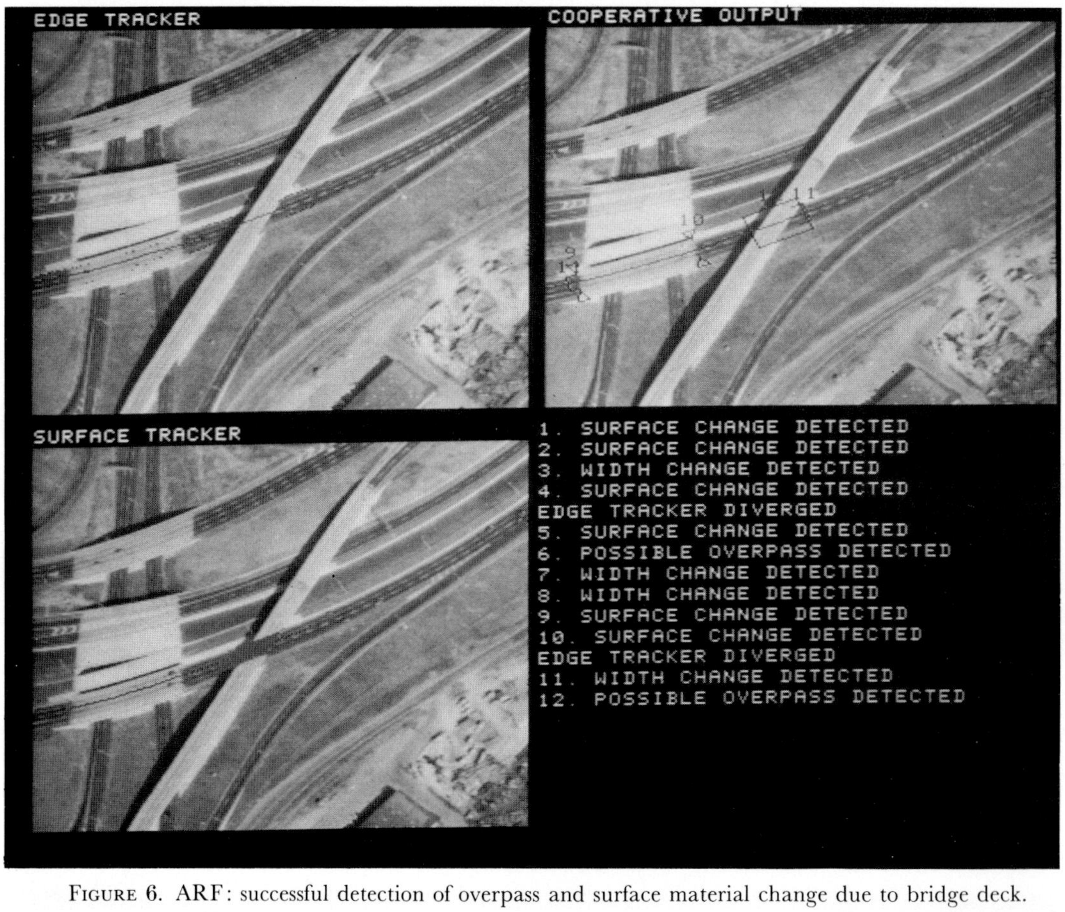

FIGURE 6. ARF: successful detection of overpass and surface material change due to bridge deck.

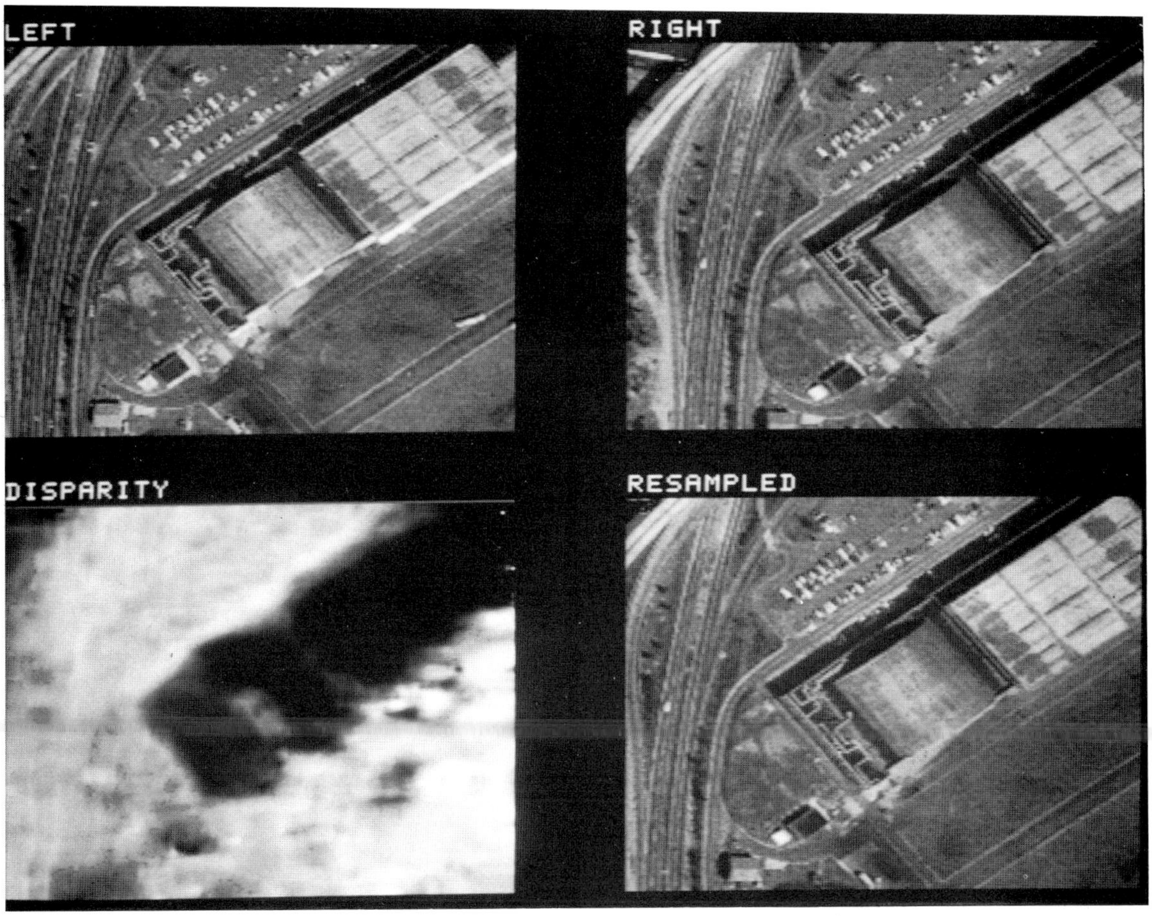

FIGURE 10. Stereo verification in airport scene analysis.

of linear features such as drainage, bridges and railroads. This view was appropriate considering the relatively low-resolution imagery, such as *Landsat* MSS, that was available at the time and in use by the image-processing community. However, we now have access to large-scale, high-resolution imagery that allows for the structural analysis of the road surface. In fact, for practical mapping applications, simply estimating the location of the road centre line is inadequate. For practical mapping applications a detailed analysis must be performed in order to maintain positional accuracy and to extract attributes of the road such as surface material, number of lanes, location of overpasses and cloverleaves.

One can categorize previous work in the area of road finder–followers into one of three major types: correlation trackers, region-based followers and edge linkers. All of these methods use a single local tracking strategy to find roads. Therefore a major problem with previous work is that if the method fails at some point it is very difficult to recover. Further, it is often difficult to automatically recognize when the tracking method has failed because the local nature of these methods assumes that the local maximum, no matter how poor, is its best guess for the position of the road.

Our approach in the ARF (a road follower) system (McKeown & Denlinger 1986) for road tracking is to use multiple cooperative methods for extracting information about road location and structure from complex aerial imagery. This system is a multilevel architecture for image analysis that allows for cooperation among low-level processes and aggregation of information by high-level analysis components. Two low-level methods have been implemented; road-surface texture correlation and road edge following. Each low-level method works independently to establish a model of the centre line of the road, and to extract various road properties such as width, surface material changes and overpasses. Intermediate-level processes monitor the state of the low-level feature-extraction methods and make evaluations concerning the success of each method. As a result of these evaluations one tracking method may be suspended due to apparent failure and restarted from the model generated by other successful trackers. Finally, a high-level module generates a symbolic description of the road in terms of various attributes of the road such as centre line, road width, surface material, overpasses and indication of potential vehicles on the road. This description is available in both map and image coordinate systems. Each tracker makes assumptions and uses road-specific knowledge in its image analysis. The surface tracker makes the assumption that the road exhibits nearly constant width and that the surface-intensity profile on the road changes either very gradually or suddenly. Abrupt changes in the surface-intensity profile may indicate surface material change, occlusions by trees or overpasses, or shadows cast by nearby buildings. The edge tracker makes the assumption that the road edges will be locally straight and antiparallel. Both trackers assume that the road exhibits a slowly changing direction and that its path can be modelled locally by a parabola. Figure 4 gives a description of the overall organization of the cooperative road tracker system.

There are several problems inherent in the use of multiple-tracking algorithms simultaneously. The major problem is one of information fusion, that is, the information generated by the two methods may give different estimates of road direction, width, etc. How can we reconcile these conflicting sources of information? A further complication is that we may not be able to determine which method is 'correct' until we proceed further down the road. One common example occurs when the surface correlator encounters a surface-material change. Therefore, it is important for the system to defer decisions and to re-evaluate decisions in the light of new evidence or situations. Both models are maintained even when we are not tracking

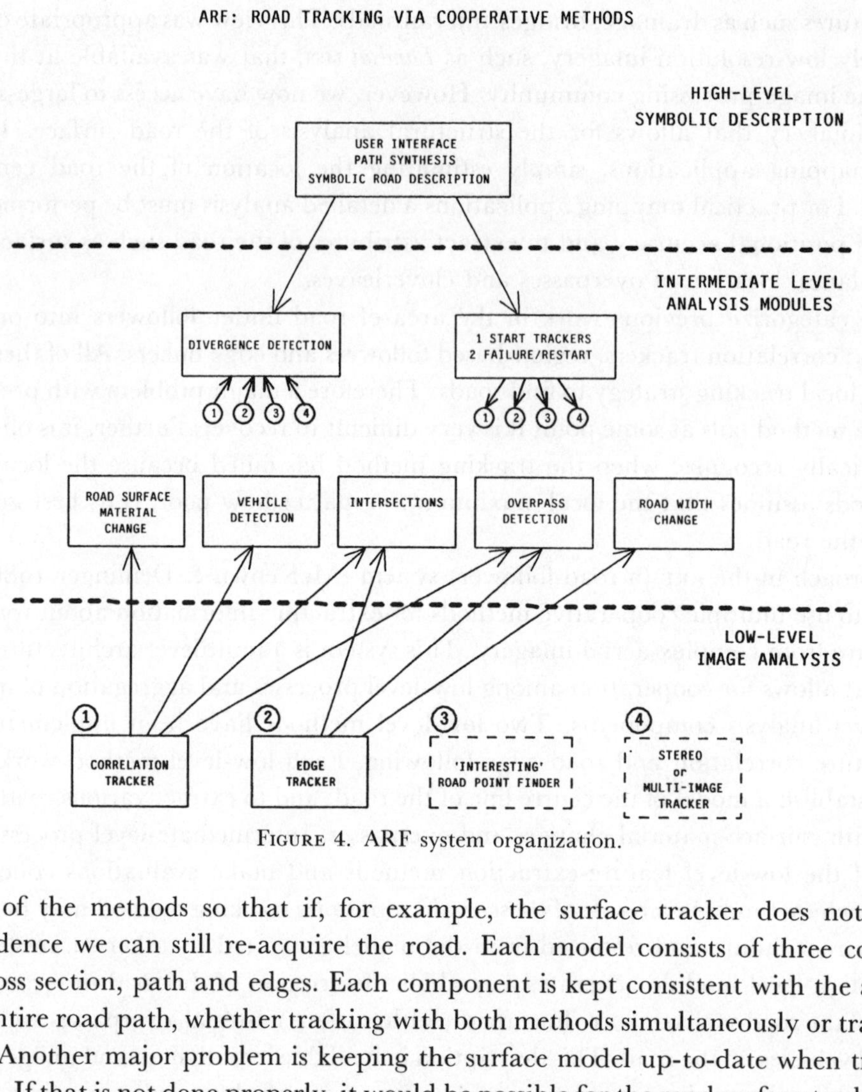

FIGURE 4. ARF system organization.

with one of the methods so that if, for example, the surface tracker does not give good correspondence we can still re-acquire the road. Each model consists of three components: surface cross section, path and edges. Each component is kept consistent with the actual road over the entire road path, whether tracking with both methods simultaneously or tracking with only one. Another major problem is keeping the surface model up-to-date when tracking the edges only. If that is not done properly, it would be possible for the road surface to change while we are tracking only the edges. In this case the road history model will not reflect the actual surface intensity. Figures 5 and 6, plate 2, show two snapshots of the road tracker running on a representative high-resolution aerial image. Note that ARF produces textual descriptions of surface-material changes, overpasses and road-width changes, annotated on the display for the user. The positions of these features are stored in both image and geographical coordinates and can be superimposed by map-to-image correspondence on other imagery of the same road. Figure 5 shows a point where the edge tracker fails, primarily because of surface-material change on a bridge-deck overpass. It is determined by the high-level evaluation module that the two trackers have diverged, that the surface-tracker model is better than the edge tracker, and therefore the edge tracker is restarted from the surface-tracker model. In figure 6 the overpass is detected as well as the bridge-deck surface change, and a width change as the road narrows because of an incoming lane.

In the following section we describe a system, SPAM, that uses spatial and structural constraints represented as production rules to interpret airport and suburban house scenes. SPAM is an example of a high-level knowledge-based interpretation system which uses several image-analysis programs to perform image segmentation and verification, but does not itself directly address image analysis. SPAM acts as a coordinator and evaluator of symbolic descriptions generated by other components performing image analysis.

5. Rule-based airport scene interpretation

SPAM, system for photo interpretation of airports by using MAPS, is an image-interpretation system. SPAM (McKeown & McDermott 1983; McKeown *et al.* 1985; McKeown & Harvey 1987) coordinates and controls image segmentation, segmentation analysis and the construction of a scene model. It provides several unique capabilities to bring map knowledge and collateral information to bear during all phases of the interpretation. These capabilities include:

(i) the use of domain-dependent spatial constraints to restrict and refine hypothesis formation during analysis;

(ii) the use of explicit camera models that allow for the projection of map information on to the image;

(iii) the use of image-independent metric models for shape, size, distance, absolute and relative position computation;

(iv) the use of multiple-image cues to verify ambiguous segmentations. Stereo pairs or overlapping image sequences can be used to extract information or to detect missing components of the model.

Figure 7 shows the overall organization of the interpretation system. SPAM maintains an internal spatial database that is composed of features extracted from imagery by various

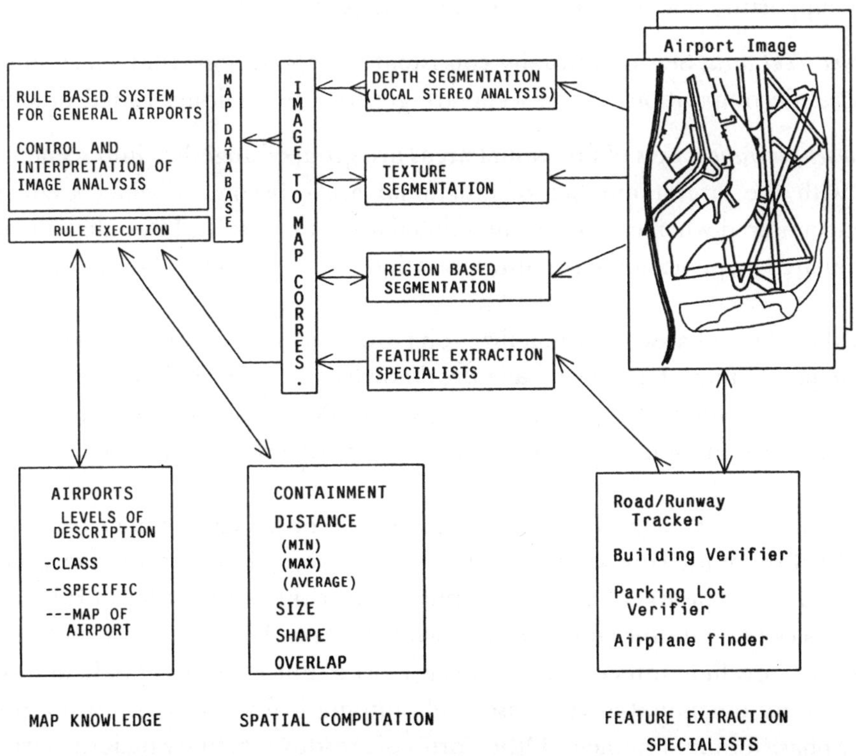

FIGURE 7. SPAM: system organization.

methods, possibly from several images, where the features are represented in terms of their geodetic position rather than their image coordinates. In fact, SPAM performs interpretation in map space which allows for a variety of knowledge such as maps and multitemporal imagery to be handled in a uniform manner. SPAM uses the following components of the MAPS system:

(i) CONCEPTMAP to retrieve shape and position models and site-specific map knowledge;

(ii) map-to-image correspondence to project models on to new imagery and image-to-map correspondence to calculate metric distances and areas;

(iii) procedures to compute spatial relations between hypotheses including containment, intersection, adjacency, closest point of approach, and subsumption.

5.1. *The SPAM architecture*

SPAM represents four types of interpretation primitives, regions, fragments, fractional areas and models. SPAM performs scene interpretation by transforming image regions into scene fragment interpretations, aggregating these fragments into consistent and compatible collections called functional areas, and selecting sets of functional areas that form models of the scene. Loosely speaking there are four phases of interpretation.

Phase 1: *region-to-fragment*. Assigns the image region data a set of fragment interpretations based solely on local properties (2D shape characteristics, texture, 3D depth, height, etc.) and knowledge about the classes of objects found in the scene.

Phase 2: *local-consistency-check*. Pair-wise tests are performed on the fragment interpretations that use spatial knowledge about the scene under consideration. The confidence of those interpretations supporting one another is incremented based on the quality of the test.

Phase 3: *functional area*. Sets of mutually consistent interpretations that share similar functions or are spatial decompositions of the scene are grouped into cliques called functional areas.

Phase 4: *model-generation*. Sets of functional areas are grouped together into scene segments. The segments with the largest number of functional areas become distinct scene models. Any conflicts encountered when combining functional areas are resolved by a default strategy, using the accumulated support for each interpretation, or by specific knowledge added by the user.

Each of these four phases operationalizes one or more types of domain knowledge. To build a SPAM system we must be able to acquire knowledge for each interpretation phase. This knowledge is represented as production rules. Over 500 productions are currently used in the airport-scene interpretation task, nearly half of which are used to evaluate and propagate local consistency.

As shown in figure 8 each phase is executed in the order given above. SPAM drives from a local, low-level set of interpretations to a high-level, more global, scene interpretation. There is a set of hard-wired productions for each phase that control the order of rule executions, the forking of processes, and other domain-independent tasks. However, this 'bottom-up' organization does not preclude interactions between phases. For example, prediction of a fragment interpretation in functional-area phase will automatically cause SPAM to re-enter local-consistency phase for that fragment. Other forms of top-down activity include stereo verification to disambiguate conflicting hypotheses in model-generation phase and linear alignment in

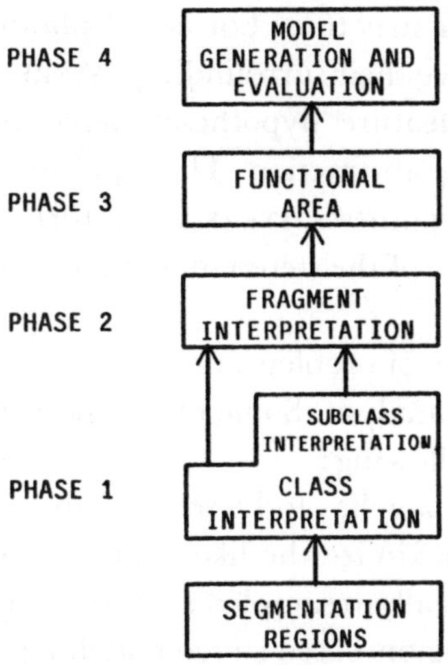

FIGURE 8. Interpretation phases in SPAM.

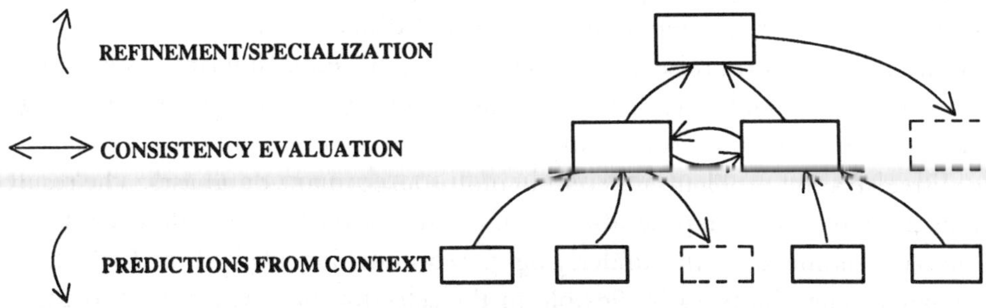

FIGURE 9. Refinement, consistency and prediction in SPAM.

region-to-fragment phase. Figure 9 shows the refinement–consistency–prediction paradigm used in SPAM within each interpretation phase. Knowledge is used to check for consistency among hypotheses, to predict missing components using context, and to create contexts based on collections of consistent hypotheses. Prediction is restrained in SPAM, in that hypotheses cannot predict missing components at their own representation level. A collection of hypotheses must combine to create a context from which a prediction can be made. These contexts are refinements or spatial aggregations in the scene. For example, a collection of mutually consistent runways and taxiways might combine to generate a runway functional area. Rules that encode knowledge that runway functional areas often contain grassy areas or tarmac may predict that certain subareas within that functional area are good candidates for finding such regions. However, an isolated runway or taxiway hypothesis cannot directly make these predictions. In SPAM the context determines the prediction. This serves to decrease the combinatorics of hypothesis generation and to allow the system to focus on those areas with strong support at each level of the interpretation.

5.2. *Stereo image analysis*

One image-analysis component of the SPAM system is a stereo-verification module, STEREOSYS (McKeown *et al.* 1986). Stereo verification refers to the verification of hypotheses about a scene by stereo analysis of the scene. Unlike stereo interpretation, stereo verification requires only coarse indications of three-dimensional structure. For aerial photography, this means coarse indications of the heights of objects above their surroundings. STEREOSYS is used within the SPAM system to confirm or refute airport feature hypotheses based on their three-dimensional structure, and operates at a high level of abstraction. The input to STEREOSYS is a segmentation region and two conflicting fragment hypotheses. STEREOSYS returns to SPAM a set of confidence factors that relate the overall confidence of the stereo match and confidence factors for each of the alternative hypotheses.

Stereo verification deals with a variety of problems that are not ordinarily present in isolated experiments with stereo matching and analysis. Some of the most interesting problems within the spatial database context are the following:

(i) an appropriate conjugate image pair has to be selected from a database of overlapping images based on criteria that would maximize the likelihood for good correspondence;

(ii) the image pairs must be dynamically resampled such that the epipolar assumption (i.e. epipolars are scan lines) used in most region-based stereo matching algorithms can be applied;

(iii) an initial coarse registration step is generally necessary because the quality of the correspondence between conjugate pairs varies greatly, in many cases the magnitude of the initial misregistration being greater than the expected disparity shift.

These requirements, in turn, raise a broad set of research issues. In terms of spatial databases the major questions are related to how an aerial image database can be used to generate automatically a useful stereo pair containing an arbitrary region, and how a stereo system can handle the misregistration problems inherent in multisource image databases. The result of this research indicate that image–map database issues in stereo verification influence the utility of such an approach as much as the underlying stereo-matching algorithm. In fact, they are intimately related. The ability to be flexible in the selection of stereo pairs provides opportunities for multitemporal, multiscale or multilook matching. Equally important is flexibility in the matching algorithm, especially with respect to assumptions that require nearly perfectly aligned conjugate images. STEREOSYS uses the following components of the MAPS system.

(i) IMAGE database to select appropriate conjugate-pair imagery bases on time, scale and flightline information.

(ii) Image-to-map correspondence to resample imagery to a common rectified projection so that epipolars align as scan lines in the image.

(iii) TERRAIN database for resampling imagery.

Figure 10, plate 3, shows an area around a hangar at National Airport in Washington, D.C. The left image contains a region of the image under analysis by SPAM. The right image was selected by STEREOSYS as the conjugate pair from the database of images maintained by MAPS. An initial registration was performed to calculate a global disparity shift between the left and right images. This disparity shift corresponds to a displacement of corresponding points in the images due to misregistration between imagery and the MAPS database. The global disparity shift is used to generate a new resampled image that accounts for inaccuracies in the MAPS database and preserves the epipolar matching constraint. The left image and the resampled image are now matched by STEREOSYS using hierarchical correlation and produce a disparity image registered to the original left image. The disparity value is one-to-one correspondence with distance, or depth, from the camera and therefore indicates relative height in vertical aerial photography. In this case, dark pixels indicate points in the disparity image that are close to the observer, light points are further away calibrated towards the ground plane of the scene.

STEREOSYS uses a statistical analysis of the disparity image to generate confidence factors that the region in the left image has low height confidence, 0.051; moderate height confidence, 0.159; and significant height confidence, 0.791. STEREOSYS generates an overall match confidence of 0.818 that confirms the hypothesis of 'hangar building'.

6. Conclusions

We have described three examples of image-analysis systems that integrate and use spatial knowledge with image analysis and interpretation techniques. Each example has been shown to rely on a spatial database of geographically referenced facts to provide information crucial to the interpretation task. Each system performs interpretation at a different level of detail, depending on the task and the overall goals of the interpretation system.

I thank the organizers of this conference, particulary Dr B. J. Conway, for a very interesting and enjoyable meeting. I also acknowledge my colleagues at the Digital Mapping Laboratory; particularly J. Denlinger, W. Harvey and J. McDermott for their contribution to the research results reported herein.

References

Barr, A. & Feigenbaum, E. A. 1981 *The Handbook of artificial intelligence* Los Altos: William Kaufmann.

McKeown, D. M. 1983 MAPS: the organization of a spatial database system using imagery, terrain and map data. In *Proc. DARPA Image Understanding Workshop, June 1983*, pp. 105–127. Arlington, Virginia: Science Applications, Inc.

McKeown, D. M. 1984*a* Digital cartography and photo interpretation from a database viewpoint. In *New applications of Databases* (ed. G. Gargarin & E. Golembe), pp. 19–42. New York: Academic Press.

McKeown, D. M. 1984*b* Knowledge-based aerial photo interpretation. *Photogrammetria* **39**, 91–123.

McKeown, D. M. 1987 The role of artificial intelligence in the integration of remotely sensed data with geographic information systems. *IEEE Trans. Geosci. Remote Sensing* **GE-25** (3), 330–348.

McKeown, D. M. & Denlinger, J. L. 1984 Map-guided feature extraction from aerial imagery. In *Proc. 2nd IEEE Workshop on Computer Vision: Representation and Control. May* 1984. Silver Spring, Maryland: IEEE Computer Society Press.

McKeown, D. M. & Denlinger, J. L. 1986 Cooperative methods for road tracking in aerial imagery. Technical Report CMU-CS-86-175, Computer Science Department, Carnegie-Mellon University, Pittsburgh.

McKeown, D. M. & Harvey, W. A. 1987 Automating knowledge acquisition for aerial image interpretation. Technical Report CMU-CS-87-102, Computer Science Department, Carnegie-Mellon University, Pittsburgh.

McKeown, D. M., Harvey, W. A. & McDermott, J. 1985 Rule based interpretation of aerial imagery. *IEEE Trans. Patt. Analys. Mach. Intell.* **PAMI-7** (5), 570–585.

McKeown, D. M. & McDermott, J. 1983 Toward expert systems for photo interpretation. In *IEEE trends and applications, May* 1983, pp. 33–39. Silver Spring, Maryland: IEEE Computer Society Press.

McKeown, D. M., McVay, C. A. & Lucas, B. D. 1986 Stereo verification in aerial image analysis. *Optical engineering* **25** (3), 333–346.

Phil. Trans. R. Soc. Lond. A **324**, 437–446 (1988)
Printed in Great Britain

Development of a knowledge-based segmentor for remotely sensed images

By A. M. Tailor†, D. G. Corr†, A. Cross‡, D. C. Hogg§, D. H. Lawrence§, D. C. Mason‡, M. Petrou‡ and R. D. Vango†

† *Systems Designers Scientific, Pembroke House, Pembroke Broadway, Camberley, Surrey GU15 3XD, U.K.*

‡ *Natural Environment Research Council Unit for Thematic Information Systems, Reading University, P.O. Box 227, Reading RG6 2AB, U.K.*

§ *University of Sussex, Falmer, Brighton BN1 9QN, U.K.*

To interpret remotely sensed data accurately, a variety of information is required about the sensor, the sensing conditions, the scene and the application. At present a good interpretation can only be achieved by a photointerpreter, unless the problem is small and can be well defined. To exploit to the full the wealth of satellite data available, efficient automatic techniques are required.

This paper describes the progress of a U.K. Alvey Man–Machine Interface Project to develop a system for the knowledge-based segmentation of multitemporal remotely sensed images. The knowledge used includes both domain dependent knowledge and knowledge contained in maps and previous segmentations. The domain-dependent knowledge includes information about the types of objects to be expected in the scene and their relations. The knowledge contained in the maps and previous segmentations could be regarded as a crude model of the situation on the ground which, although probably only partly correct, should still greatly facilitate segmentation.

A major objective is to increase classification accuracy in images of the type used in the environmental sciences. A further use could be to update the existing map data as a result of segmentation.

Introduction

The volume of quantitative data from remote-sensing platforms has increased considerably in the last decade. This has resulted in a corresponding increase in the scope and magnitude of the problems to which the images are applied. Present-day problems that use remotely sensed images cover applications as widely varying as urban land use studies and crop inventories (see, for example, Landgrebe (1981) for typical applications).

Traditionally, interpretation of remotely sensed data is performed by a photointerpreter using information in the image, together with his expertise and any available data, such as maps and ground truth. A full interpretation of the image is still best performed this way, although it is a lengthy process. There is a need for efficient machine-implemented analyses of the data.

The automatic techniques used for image interpretation are traditionally those of pattern recognition. The stages of a typical pattern-recognition system are shown in figure 1. The input to such a system for remote sensing is a digital image. The image is first preprocessed as necessary; this may not only include standard corrections, for example, registration to a reference grid or correction to reduce sensor-induced noise or distortions, but may also include

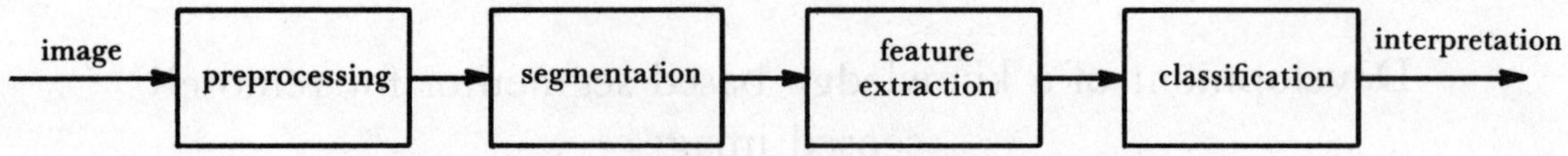

FIGURE 1. Block diagram of a traditional pattern-recognition system.

image transformations to enhance or suppress specific image features. The preprocessed image is then partitioned into segments, or regions, at the segmentation stage. Features, or properties, of the regions are extracted which describe each region uniquely and enable the correct class to be assigned to the region at the classification stage. A full interpretation of the scene can then be performed.

Most automatic analyses of remotely sensed imagery still rely only on the supervised spectral analysis of pixels on an individual basis. This process omits the automatic segmentation stage: a photointerpreter trains the classifier by selecting typical examples of the regions of interest. The spectral properties of the example regions are used as identifiers of the relevant class. Each new pixel is assigned to the class whose spectral properties are most similar to those of the pixel. Extensive training is required and the method is generally only satisfactory for highly constrained applications, where the spectral properties of the specific regions are sufficient for the problem to be adequately defined.

More recently, unsupervised methods have become more common. Here, no *a priori* assumptions are made about the classes of regions in the data and automatic techniques are used to divide the image. In addition, methods that use information other than the spectral values of pixels are being used. These considerations will be discussed in the sections below.

SEGMENTATION

Definition of segmentation

There is no absolute definition of segmentation in the literature. Segmentation may be said to be a partitioning of an image into segments, each of which is a group of adjacent pixels defined by properties which lie within a certain range. This reduces the problem to finding properties which, when their values are limited, define similarity within spatially connected areas of the image.

Segmentation is application dependent in that different levels of partition will be required by different applications. For example, an agriculturist requires segmentation to the individual field level, whereas a forester is content with a larger agricultural category.

Segmentation is also sensor dependent: two images of the same scene taken at different spatial resolutions, or that use different spectral wavebands, will produce a different segmentation. In addition, variations in the sensing conditions across an image and between images taken at different times will affect the segmentation: not only the values, but also the relevant properties required for isolation of an object may vary.

Goals of segmentation

The purpose of a segmentation is to create a low-level representation of the image which retains as much information as is necessary for classification, while presenting that information in a straightforward manner. To this end, there are two principal goals of a system for segmentation, namely: to produce a reliable segmentation; to improve classification accuracy.

For investigation of general segmentation methods, the influences of the application, the sensor and the day-to-day variations must be reduced. The dependence on the application cannot be eliminated, but segmentation to some arbitrary level can be attempted. However, the techniques must be parametrized to cater for all input image dependencies.

Segmentation techniques

Segmentation techniques may be region-based or edge-based. Region-based techniques locate regions by using a measure of similarity between the given pixel properties. Edge-based techniques locate the region boundaries by detection of large discontinuities in the given properties. All such techniques are data-driven in that they rely on the properties in the input image to define the segments.

Region-based techniques do not cope well with thin regions and the boundaries are not normally located accurately. Examples of such techniques used for remotely sensed data are clustering (see, for example, Townshend & Justice 1980; Seddon & Hunt 1985) and split-and-merge (see, for example, Cross & Mason 1985). Clustering iteratively merges pixels until the similarity criterion is satisfied for each region. Split-and-merge techniques successively split and merge blocks of adjacent pixels to satisfy the criterion.

Edge detection techniques are principally concerned with locating points of large gradient and linking these points into connected boundaries (see, for example, Nevatia & Babu 1979). Consequently, such techniques are more successful in locating the region boundaries and thin regions, provided that the contrast is strong enough.

Errors in segmentation

None of the standard segmentation techniques is adequate for the segmentation of complex images from aircraft or satellite. Even if the technique performs to the best of its ability, segmentation errors and ambiguities still occur. These errors manifest themselves as over-segmentation, in break-up of a region, and undersegmentation, in either full or part loss of a region.

These errors may be due to practical issues, for example the sensing conditions, but the principal source of error is the data-driven aspect of the techniques. There is generally insufficient information in the image itself to discriminate completely between regions. Even if the information is in the image, the exact properties and values to use are not known *a priori*. Recourse to information outside that in the input image is necessary.

Exploiting knowledge

Knowledge available

For more accurate segmentation, the information in the image, given by the spectral, spatial and temporal variations of the pixels, can be supplemented by *a priori* knowledge. This knowledge, which imposes constraints on the regions to be expected in the data, is external to the image.

The *a priori* knowledge can take the form of either additional data sets or domain data. The additional data sets may be maps or other sensor data of the same scene. However, some aspects may differ from those of the original data, for example the data representation or the spatial resolution. The domain data, which represent the expertise of an analyst, are generally in the form of conditions on the regions in the data; for example, fields generally have straight boundaries.

The available knowledge may act on the properties of regions at the segmentation level or on the classes of regions at the classification level. The spatial information on the properties may be two- or three-dimensional.

Techniques for incorporating knowledge

The standard region and edge-segmentation techniques mentioned previously generally only use limited spectral and spatial information from the image. Techniques that exploit other forms of image information include relaxation, which uses local relations between pixels (see, for example, Peleg 1984), and signature extension methods, which use the temporal signature of objects to aid classification (see, for example, Henderson 1976).

The exploitation of information external to the image requires more sophisticated techniques for data-integration and imposition of constraints. A review of such techniques can be found in Tailor *et al.* (1986). In particular, image understanding systems (Matsuyama 1986) may be built up to improve aspects of the interpretation process. Ideally, the knowledge would be used as a model of the state on the ground, for example, the system of McKeown *et al.* (1985) for the interpretation of airport scenes in aircraft data. However, the complex nature of the data means that the knowledge is rarely available in the right form or is not accurate enough for it to be used this way. More often, the knowledge is used as rules in a rule-based system to impose constraints on the regions, for example, the system of Nazif and Levine (1984) for segmentation of aerial imagery. Depending on the nature of the information, most systems combine the modelling and the rule-based approaches.

The remaining sections of the paper are concerned with a knowledge-based system which we have developed for segmentation of multitemporal, high-altitude aircraft and satellite data. The emphasis is on general segmentation techniques, as in the Nazif & Levine (1984) system, rather than on interpretation of more specific information, as in Goldberg *et al.* (1985) and McKeown *et al.* (1985). However, compared with the system of Nazif & Levine (1984), we include a modelling, or goal-directed, approach to supplement the data-driven approach, together with the use of temporal information.

Knowledge-based segmentation system

The preliminary system for knowledge-based segmentation developed in this project uses map data, together with domain data in the form of rules, to supplement the information in a multitemporal aircraft data set. Segmentation is to the lower order of segment, e.g. a field, rather than the higher order of segment, e.g. agriculture.

The rules in the rulebase have been designed to use as little knowledge of the classes of the regions in the data as possible, other than that in the map. The classification information that is known *a priori* is generally either specific to an application or depends on the sensing conditions. As the system improves it is the intention to add classification information as necessary.

The knowledge used is currently all two dimensional; no elevation data are included.

The approach to segmentation is a combination of the data-driven and the goal-directed approaches, the latter being evident due to the use of the map as a crude model of the situation on the ground.

The basic system is modular and is shown in figure 2. An initial segmentation of the current image is formed by imposing lines, for example roads, from the map onto a region segmentation of the image. An edge segmentation of the image is produced to supplement the initial segmentation. Features of the regions are calculated which are used to satisfy constraints in the rulebase. These features include the uniformity of the region, the contrast between the region and a neighbour, and the support to the region boundary from edge and line pixels.

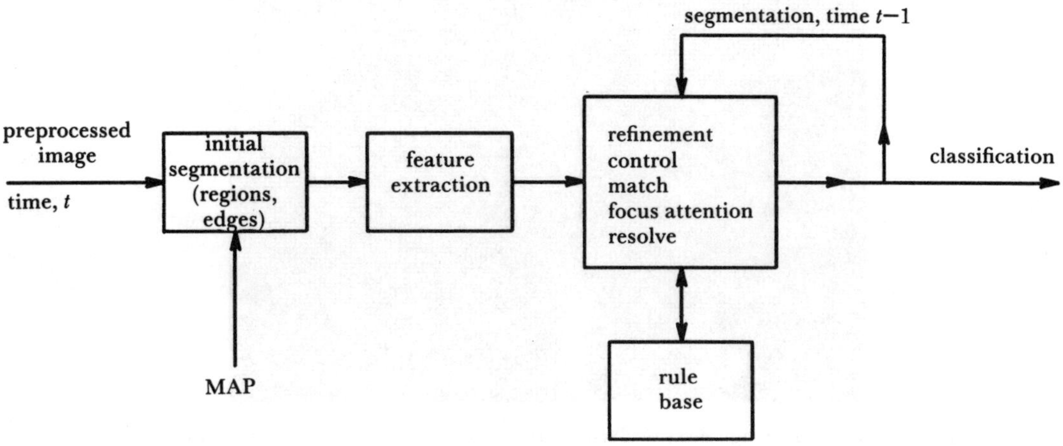

FIGURE 2. Block diagram of the knowledge-based segmentation system.

The initial segmentation is refined, based on the evidence from the region features and the constraints imposed on the regions in the rulebase. At this stage, the segmentation of the previous image in the time sequence, if available, is matched to the initial segmentation to assess temporal consistency. A measure of confidence is assigned to each region based on the compatibility of the evidence. Attention is focused onto ambiguity in the segmentation, as represented by low-confidence regions. If the evidence strongly supports a change, a region may be modified by merging with a neighbour, splitting, or adjusting a portion of the boundary.

After processing all the images in the time sequence, the segmentation may be classified.

DATA

The system has been tested on multitemporal image data from the NERC simulated thematic mapper aircraft data. Each image has 11 spectral channels and spatial resolution of 10 m. The data were transformed by using a principal component transform to decorrelate the spectral information.

Figure 3 shows a 64×64 pixel, first principal component section of one image in a three-image time sequence. The image consists of a combination of unstructured, natural fenland in the top-right corner and structured fields with well-defined boundaries. There is little elevation.

Figure 4 shows an area of digitized Ordnance Survey 1:10000 map data which corresponds to the image. Each of the lines in the map is 1 pixel wide. The two lines running from left to right across the map are roads; the incomplete lines running into the fenland are footpaths; the remainder are field boundaries.

A hand segmentation has been produced for each image, which is used as a reference for validation of the segmentation. Figure 5 shows the hand segmentation into regions corresponding to figure 3. There is a corresponding hand segmentation into lines which includes the

FIGURE 3. Aircraft image data of Barton Broads, spring 1986. The figure shows a first principal component, 64 × 64 pixel image at resolution of 10 m.

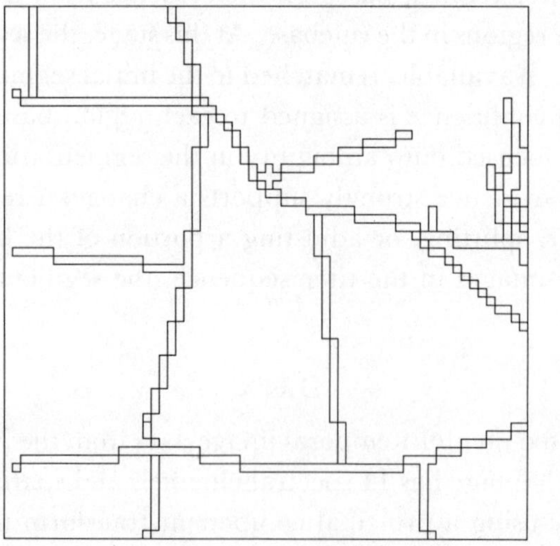

FIGURE 4. Digital Ordnance Survey 1:10000 map data corresponding to the area of figure 3.

thin regions as suggested by the map. The hand segmentation has been drawn up by a photointerpreter using information on the identity of regions from the image, the map and the ground truth.

The image, the map and the hand segmentation have each been registered to the British National Grid.

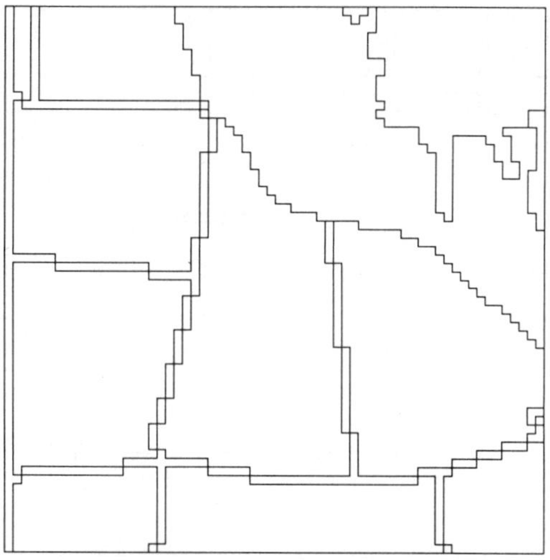

FIGURE 5. Hand segmentation of the scene into regions.

TESTS

The system was tested by using the image and map data described above. Various configurations of the system were tested to assess the usefulness of the knowledge. For example, the system was run with and without the map data.

In addition, a segmentation by a clustering procedure was performed on the last image in the time sequence for comparison. Both segmentations were validated against the hand segmentation.

To assess the ability of the system to improve classification, the segmentation was classified by using a region classifier. The classifier assigns a region to the class whose spectral characteristics are most similar to those of the region. The classes are defined by a photointerpreter labelling the regions in the hand segmentation. The classification of the segmentation was compared with that obtained by a standard per-pixel classifier.

The percentage of correctly classified pixels was calculated for both the region and the per-pixel classifications.

RESULTS

Figure 6 shows the result of the knowledge-based segmentation of the image in figure 3. In general there is good correspondence between the regions and those of the hand segmentation in figure 5.

The boundaries of regions are not detected well: they are often misplaced, doubled up or represented by a number of small regions. This is due to a number of factors.

Firstly, the line information, after being imposed on the initial segmentation, is allowed to change at the refinement stage. This is consistent with the assumption that the information may be incorrect. The condition for line accuracy is however too strict, resulting in the field boundaries in the centre of the image being modified.

Double boundaries result in general from lack of good registration between the image and map data: both strong edges and map lines may reinforce a boundary, but at different positions.

The break up of a boundary into small regions is due to the large number of high-contrast edges present. A similar effect occurs in high variance or textured regions such as woods or

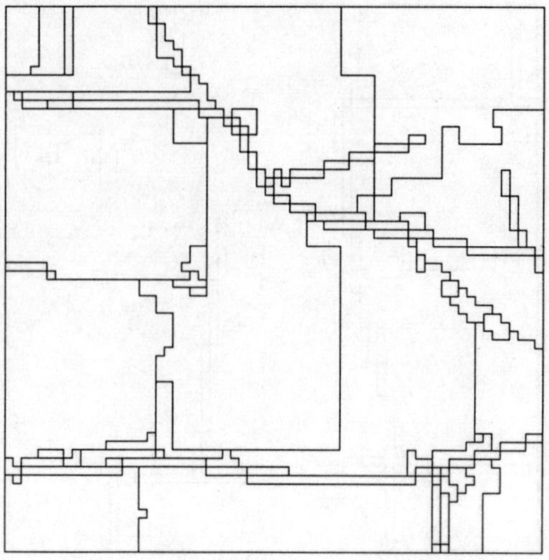

FIGURE 6. The result of knowledge-based segmentation of the scene.

urban areas. The system at present has no provision for combining these small regions into a consistent interpretation. However, this is principally a problem for the higher level stages of classification and interpretation.

Undersegmentation is a problem in the top-right corner of the image where the spectral values of the pixels are insufficient for discrimination of the two regions in the fen area. This is due to the segmentation being performed on only the first principal component of the data. The area is too uniform in the one component for any evidence to be gathered for splitting the resultant region.

Figure 7 shows the result of the segmentation by clustering of the image. Oversegmentation of high-variance regions is also a problem here, not only on the boundaries, but also interior to regions. In addition, undersegmentation is much more evident than in the knowledge-based segmentation; there is no means by which the clustering can divide similarly valued pixels. On this basis, the knowledge-based segmentation performance is superior to that of the clustering.

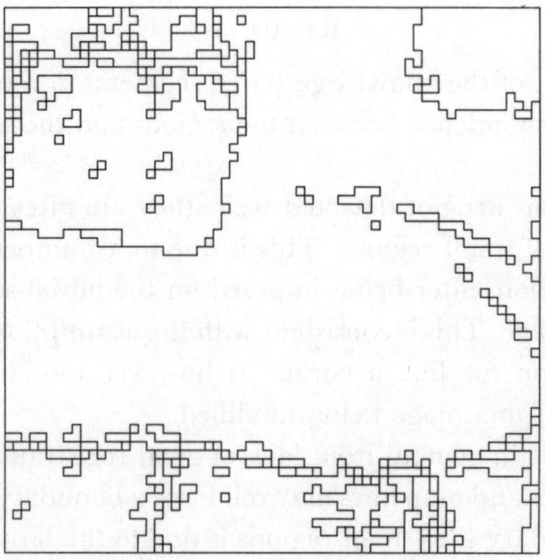

FIGURE 7. Segmentation of the scene by a standard clustering procedure.

The classification of the knowledge-based segmentation of the image gave 81.4 % of pixels correctly classified.

Many of the segmentation ambiguities due to oversegmentation at the boundaries and in textured regions were resolved at the classification stage, because of the regions having spectral values sufficiently similar. However, the undersegmented area in the fen, because it was made up of pixels from two different distributions, was classified as neither class. This contributed greatly to the classification error.

The presence of the roads from the map in the segmentation caused the majority of the remaining classification error. These line features were not included in the region hand segmentation, but are present in the line hand segmentation and are treated separately.

The accuracy of the per-pixel classifier was 86.2 %. The greater accuracy was due to the two regions in the fen area being correctly classified when examined on a pixel-by-pixel basis with more than one spectral component. This classifier, however, suffers from high variance in the image in a similar way to the segmentation: many small regions result. The manifestation of the small regions at this level of the interpretation process is much more difficult to amend.

Discussion

The preliminary system for knowledge-based segmentation described here compares favourably with the standard automatic techniques. However, there have been a number of inadequacies isolated in the knowledge-based system.

The initial segmentation is not adequate for successful refinement to take place. Oversegmentation at this stage in high variance regions causes problems: a separate system module is required for dealing with these areas. In addition, the initial region and edge segmentation procedures used cause many spurious features in noisy areas. This is due partly to the inferior techniques and partly to the lack of a full multispectral capability.

The refinement stage is deficient in one major area: the treatment of regions or their boundaries which have resulted from the imposition of map lines. Greater emphasis on these areas is required to avoid losing the information. This should include a mutual enforcement of map lines and image edges.

It is planned to improve and extend the system in the light of the results to produce a demonstration operational system. This system will be designed principally for the land use applications of environmental and agricultural monitoring. However, we envisage that it will have general applicability to the remote-sensing community. This is due to the modularity of the system which will allow its adaptation to other sources of knowledge or other requirements. This final system will also provide explicit information on map updates and image change detection. Future tests will include the processing of other sensor data, such as that from the French *SPOT* satellite and synthetic-aperture radar data.

The final stages of the current project will include studies of the hardware required by a full operational system, the augmentation of the knowledge-base to include classification information and pointers to potential areas of research.

We thank Nigel Brown and David Norris of the Institute of Terrestrial Ecology who have been responsible for digitizing the Ordnance Survey Maps.

References

Cross, A. & Mason, D. C. 1985 Proc. Int. Conf. of the Remote Sensing Society and the Center for Earth Resources Management on Advanced Technology for Monitoring and Processing Global Environmental Data, London, 9–12 September 1985.
Goldberg, M., Goodenough, D. G., Alvo, M. & Karam, G. M. 1985 *Proc. IEEE* **73**, 1054–1063.
Henderson, R. G. 1976 *IEEE Trans.* **GE 14**, 34–37.
Landgrebe, D. A. 1981 *Proc. IEEE* **69**, 628–642.
Matsuyama, T. 1986 In *Proc. Int. Geoscience and Remote Sensing Symp, Zürich, 8–11 September 1986*, pp. 277–282. ESA Publications Division.
McKeown, D. M., Harvey, W. A. & McDermott, J. 1985 *IEEE Trans.* **PAMI 7**, 570–585.
Nazif, A. M. & Levine, M. D. 1984) *IEEE Trans.* **PAMI 6**, 555–577.
Nevatia, R. & Babu, K. R. 1979 In *Proc. 6th Int. Joint Conf. on Artificial Intelligence, Tokyo, 20–23 August 1979*, vol. 6. pp. 639–641.
Peleg, S. 1984 *Comput. Vision Graph. Image Process.* **25**, 122–130.
Seddon, A. M. & Hunt, G. E. 1985 *Int. J. Remote Sensing* **6**, 717–731.
Tailor, A. M., Cross, A., Hogg, D. C. & Mason, D. C. 1986 *Image Vis. Comput.* **4**, 67–84.
Townshend, J. R. G. & Justice, C. O. 1980 *Int. J. Remote Sensing* **1**, 105–120.

Discussion

D. Lane (*Intelligent Automation Laboratory, Department of Electrical and Electronic Engineering, Heriot-Watt University, Edinburgh, U.K.*). Would Dr Tailor briefly describe the hardware and software environments that she is using for this work, and comment upon the advantages and disadvantages that have resulted.

A. M. Tailor. The system has been implemented in POP 11, a general-purpose programming language, integrated into the POPLOG environment. POPLOG has been developed at Sussex University and is marketed commercially by Systems Designers. The system is running both in a VAX environment and on a SUN 3 workstation.

Major advantages have been found using POP 11; namely the ease of development and flexibility in implementation. However, for an operational system, both some recoding into a language such as PASCAL or C, and use of special-purpose hardware would be required, because of the current long execution times.

Phil. Trans. R. Soc. Lond. A **324**, 447–456 (1988)
Printed in Great Britain

A knowledge-based approach for evaluating forestry-map congruency with remotely sensed imagery

By M. Goldberg[1], D. G. Goodenough[2] and G. Plunkett[2]

[1] *Department of Electrical Engineering, University of Ottawa, Ottawa, Canada K1N 6N5*
[2] *Canada Centre for Remote Sensing, Ottawa, Canada K1A 0Y7*

[Plate 1]

Simple algorithmic methods for integrating remotely sensed data with existing cartographic databases do not provide satisfactory results. One of the main difficulties is in reconciling spatial differences between the two data sources. The differences are due to such diverse factors as temporal variations, spatial errors in the map data, and topographic effects in the remotely sensed data. To investigate the extent of this data-integration problem, a map–image congruency evaluation (MICE) knowledge-based system was developed which performs three distinct operations: (i) pre-processing for uniform representation of both the image and cartographic datasets; (ii) spatial reasoning on the data with the MICE system; (iii) presentation of a congruency evaluation map. Results are presented for a forested area in British Columbia, where *Landsat* multispectral scanner data are integrated with a provincial forest-cover map.

1. Introduction

Geographic Information Systems (GIS) are computer-based systems for handling map-like information. These can be used to generate maps, to retrieve statistical information and to update the geographic databases, to name but a few functions (Marble 1984). An example of one such system is the GIS used and operated by the Inventory Branch of the British Columbia Ministry of Forests and Lands (BCMOFL), which covers an area of approximately 52×10^{10} m^2 and which must be updated annually with respect to forest depletion (Hegyi & Sallaway 1983). Satellite-based, remotely sensed imagery, because of its large and regular coverage, has the potential for being used for this updating. The process of extracting information from the remotely sensed imagery and its incorporation into the GIS is called 'data integration'. A serious problem that has been noted is the spatial incongruencies between the map and the image data (Goodenough 1988).

Existing databases represent large investments and there can be reluctance to make changes or corrections even when it is recognized that there are errors. Often, the new data elements to be integrated are moved to a location that 'fits' into the map, rather than making extensive modifications to existing elements of the database. This implies that even though the remotely sensed imagery is rectified to the same projection coordinates as the cartographic data, random, nonlinear spatial discrepancies may still exist. This has important implications if automated techniques are to be used for updating the GIS. Somehow, information related to the spatial discrepancies must be acquired.

Algorithmic methods for correlating the map and image data have met with only limited success (Parsons 1984; Billingsley 1982). The essential reason is that features in images do not correlate well with features on maps. However, as maps are produced from imagery (photographs) by experts, a knowledge-based approach may prove to be more productive.

The remainder of the paper is divided as follows. Some background information about photogrammetry, expert systems, and the programming tools used for this research are presented in §2. The map–image congruency problem and a knowledge-based approach are presented in §3. Finally, some examples illustrating the performance of the expert system constructed are provided in §4.

2. Background

2.1. *Photogrammetry*

Photogrammetry is the science of the tranformation of photographs into maps. Most GIS systems today contain map data that were collected, analysed and input by photogrammetric techniques. Older paper maps, that were generated manually, may be digitized and put into a GIS system (Boyle 1980). To understand the nature of errors in a GIS, it is necessary to review some of these photogrammetric techniques.

Photogrammetric data acquisition usually requires taking air photographs with precise altitude, attitude, and positional information. The map is usually then generated by manually tracing desired features from the air-photograph mosaic. The output of the tracing may go directly into a GIS (digitization) or on to hardcopy (scribing).

There are a number of differences between a map and a photograph, which must be taken into account when analysing air photographs. These differences include:

(*a*) maps are drawn to a predetermined scale;

(*b*) maps display only selected features, depending on their purpose;

(*c*) maps emphasize certain selected features;

(*d*) maps display features by using standard symbols;

(*e*) maps are generalized (i.e. some detail is lost);

(*f*) maps are lettered, titled and labelled;

(*g*) maps are produced based on some standard projection.

When tracing an air-photograph mosaic, the photograph interpreter must use considerable judgement in identifying the features of interest. Some of the clues used in feature identification include the following:

(*a*) natural features often have irregular shapes, whereas man-made features have regular outlines;

(*b*) shadows in the photograph give an indication of the height (or depth) of a feature;

(*c*) adjacency relations give a clue to feature identification (e.g. a road crossing a river would be a bridge);

(*d*) stereo viewing allows one to mentally generate a three-dimensional understanding from two two-dimensional images.

Many of these feature identification clues are so obvious that they are often 'second nature'. However, each of these clues could add essential information to the feature identification problem in an expert system.

2.2. *Expert systems*

'Artificial intelligence (AI) is the study of ideas that enable computers to be intelligent' (Winston 1984, p. 1). AI encompasses a number of domains including natural language understanding, problem solving, robotics and vision, and expert or knowledge-based systems. All expert systems have three common features (Hayes-Roth *et al.* 1983):

(*a*) a human interaction capability to allow human input and machine output;

(*b*) an inference engine to control the deductive reasoning;

(*c*) a knowledge base to store the domain-specific knowledge used by the inference engine.

The architecture and techniques used in expert systems are varied and often complex.

Several rule-based systems for machine perception in remote sensing applications have been developed. Glicksman & Mackworth (1982) researched the use of multiple information sources for image understanding in the MISSEE system. McKeown (1984) and McKeown *et al.* (1985), performed map-assisted photograph interpretation in the MAPS/SPAM system. Plunkett (1986) and Plunkett *et al.* (1986) examined the spatial congruency of forest cover maps and *Landsat* MSS (multispectral scanner) imagery in the MICE system.

2.3. *RESHELL*

One approach to reducing the development cost of an expert system is to use a package called an expert-system shell. A shell contains most of the elements of an expert system except the domain knowledge. The knowledge engineer can easily add rules to the knowledge base of the shell and thus develop an expert system more rapidly. RESHELL (Goldberg *et al.* 1985; Goodenough *et al.* 1987) is an expert-system shell written in Logicware's version of PROLOG, known as MPROLOG.

The architecture of RESHELL is given in figure 1. RESHELL is very modular in the sense that it supports development of a multi-expert system, with individual experts organized hierarchically using blackboards for a communication medium. Each expert in the hierarchy is responsible to a single manager, so that control and communication flow between different levels of the hierarchy rather than across a level. The highest-level expert sets broad goals for the next level of command. The lowest level corresponds to the image-processing algorithms coded in FORTRAN. The two major parts of an expert system are the knowledge base and the inference engine. The knowledge base consists of facts and rules. Facts are the basic information regarding a problem, whereas rules are applied in solving a problem. Rules are in the form of production rules: 'if CONDITION then CONCLUSION'. The inference engine is the scheduler or control mechanism. There are two control strategies that can be followed, namely bottom-up (backward-chaining), and top-down (forward-chaining). Backward-chaining refers to the conclusion being true and finding rules leading to the conditions being true, working backwards from the final goal to the initial state. In forward-chaining, the condition is assumed true and the conclusions are deduced which can be used in other rules leading to the final goal.

Other features presented in RESHELL include a knowledge-acquisition system, a justifier to explain the reasoning of a program, methods for treating uncertainty in knowledge and data, a semantic relation or frames database, as well as an external communications method. The frames database stores data in a manner that preserves descriptive semantic relations between

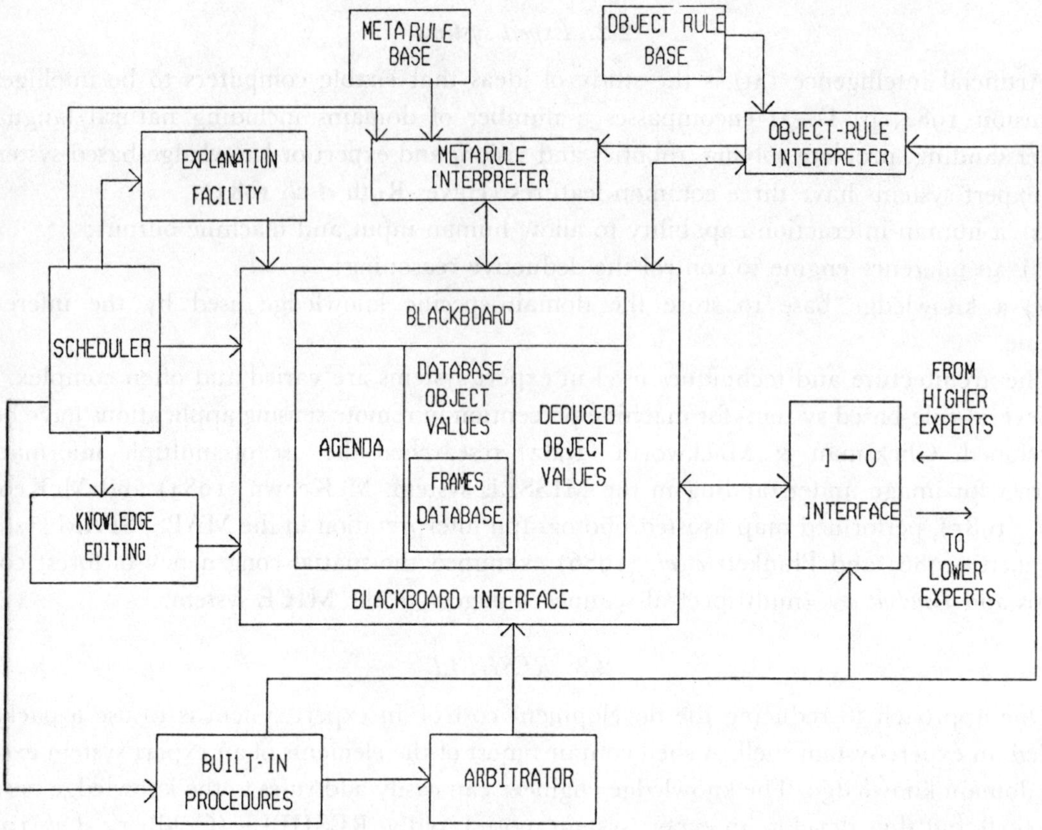

FIGURE 1. The architecture of RESHELL.

objects. The interface to the FORTRAN image-processing algorithms consists of an expert, called the LTI (LDIAS Task Interface), that is responsible for acquiring all the necessary knowledge to execute the FORTRAN code in batch mode, when activated to do so by a higher-level expert. There are two levels of rules: metarules and object rules. Object rules are inference rules that manipulate objects to deduce or prove goals. The object-rule interpreter makes inferences using object rules. Metarules control the metalevel interpreter to select the best appropriate path for the next stage of a solution by manipulating sets of object rules.

The blackboard stores goals from higher-level experts, current object values, the agenda created by the metarule interpreter, as well as the intermediate results. The responsibilities of the scheduler include: receiving messages from the I–O interface; invoking the metarule interpreter; evaluating action procedures in the agenda of the blackboard; and calling built-in procedures when required. The data interface is used for communication among different experts at different levels of the hierarchy. Finally, the arbitrator resolves conflicting results if multiple strategies and/or approaches are used for problem solution.

3. Comparing maps and images

3.1. *The map–image congruency problem*

In the map-making–updating process, photograph-interpreters typically analyse aerial photographs, decide on the classification of the different objects in the photograph, and then transcribe the classification and location of these objects on to a map or directly into a

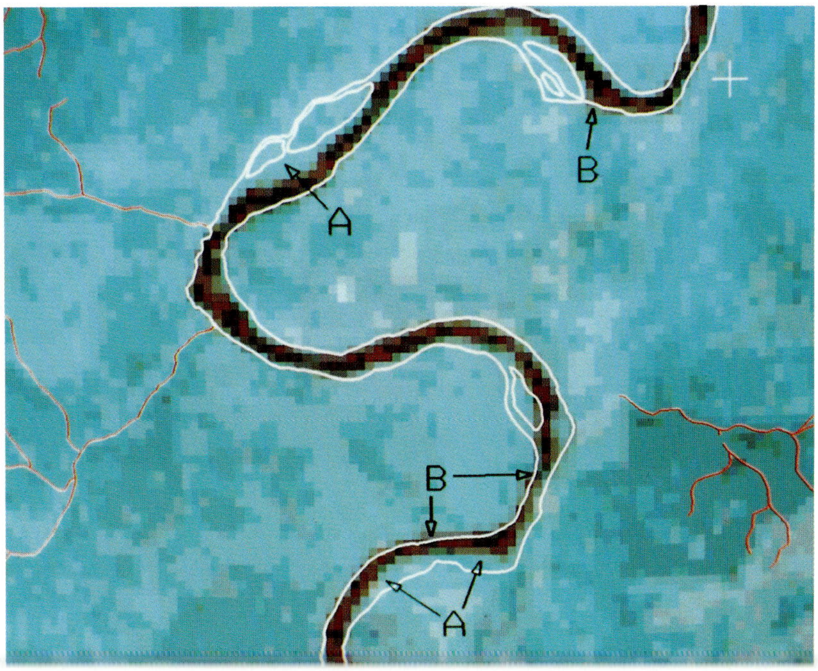

FIGURE 2. River mismatch. Area A indicates land pixels within the map river boundary; area B indicates water pixels outside the map river boundary.

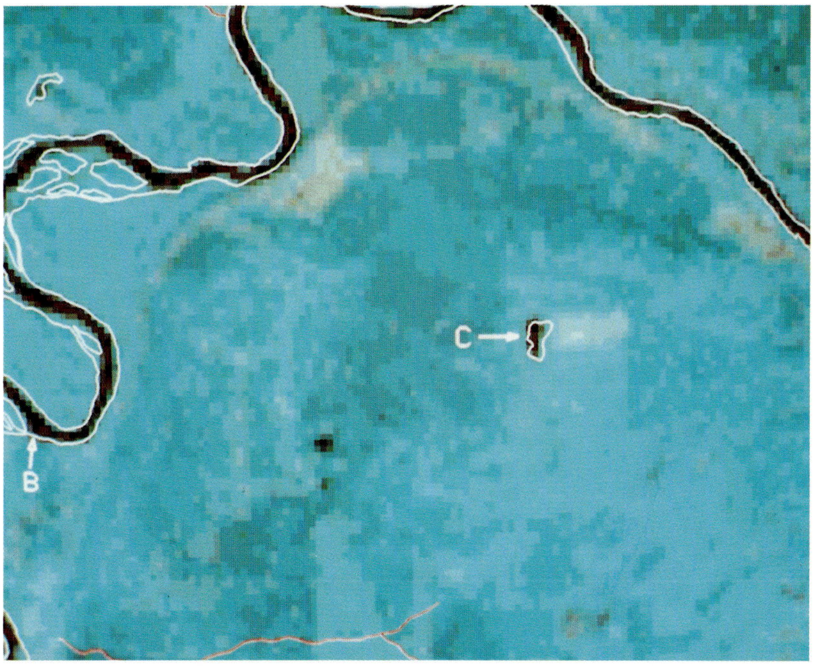

FIGURE 3. Lake mismatch. Area C indicates misalignment of the lake in the map and *Landsat* MSS image.

geographic information system (GIS) (Zarzychi & Allam 1982). The map is only an approximation to the real world; furthermore, this map-making procedure is prone to human error. A further complication is that the world land-mass is a changing entity. For example, rivers meander, forests burn or are cut, and housing subdivisions and roads are built. Cartographic data, on the other hand, are relatively static and are only updated periodically to reflect the changing world.

The integration problem occurs when some area in a remote-sensing image, which is usually depicted as a polygon, is to be placed in a GIS. If the image polygon is placed directly into the GIS database, then database corruption will occur, if there is not perfect spatial juxtaposition of the image polygon and the neighbouring map polygons. The current method of performing this integration is for a human operator to move the image polygon into the best-fitting location in the map and then place it in the GIS database.

Examples depicting the map–image misregistration problem are given in figures 2 and 3, plate 1. These figures depict a geocoded *Landsat* MSS image overlaid with the hydrography level of the corresponding map. Area A of figure 2 contains pixels that are clearly land in the image, but are categorized as river in the map, whereas in area B the opposite phenomenon has occurred; pixels that correspond to water in the image are categorized as land in the map. Figure 3 is a second example, where the lake in the map appears to be in a slightly different location than the lake in the image (area C).

3.2. *The photograph-interpreter approach*

In dealing with map–image misregistration, an expert photograph-interpreter would be called on to make many judgement calls. Both formal and heuristic rules would be used in making decisions on how to reconcile the two data sources and finally on how to update the map. An expert system paradigm seems to be an appropriate framework for attacking this complex problem. The map–image congruency evaluation (MICE) knowledge-based system (KBS) uses such a paradigm.

Another question that must be answered before selecting a strategy is: how does one know that the map and the image match? A person would look at the map and then at the image, find the corresponding structures, measure their position from some datum, and then report the congruency or discrepancy for the structure. The strategy selected for this rule-based system is not like the human approach. The MICE system performs the same basic operations. These are: (1) pre-process the map and the image to the same spatial datum and symbolic representation; (2) locate corresponding structures (segments); and (3) report on the spatial congruency of the corresponding structures.

There are many types of spatial incongruency. The incongruencies can be random or systematic, local or global, and large or small. The current strategy is simply to report the results and not to try to fix the misregistrations automatically. The evaluation is left to the user.

3.3. *Symbolic representation of maps and images*

One of the first questions that had to be addressed in the design of MICE was how to represent the two disparate data types, map and image data, in a knowledge-based system. This is the iconic-to-symbolic gap problem that is being researched for machine perception systems (Eshera & Fu 1986). Map data containing dots, lines and areas are usually defined in a coordinate reference system in terms of points, vectors and polygons. Image data, on the other hand, are stored by a spatially indexed technique called raster or grid format. It would be extremely unwieldy to attempt to store and process these data in their native form in a knowledge-based system, as the format, data type and resolution of the data are different. Also, the system does not necessarily make decisions based on the data, but rather on various attributes derived from the data. Thus, the uniform method of data representation selected was to pre-process the map and image data into segments and to generate various symbolic segment attributes that can then be used by the rule-based stage.

The next question that needs to be answered is: which attributes of the data are required for congruency evaluation? This question is also not easy to answer, because the image has spectral attributes that are not available in the map data. The image segments' spectral attributes are required, so that the image segments can be spectrally classified. The map and image spatial attributes can be calculated relative to the same reference grid, so that the attribute values of the map and image can be compared and manipulated in a symbolic fashion.

Note that: (1) both the map and the image have been processed to the same spatial resolution and thus can be compared on a pixel basis; (2) this means that the attributes are pixel-size-invariant. Thus the pixels can be any size as long as the map data and image data are pre-processed to the *same* resolution. The spatial attributes selected for use by MICE, that are common to both the map and the image, are as follows:

(*a*) the location of the pixels in the map or image segment, based on some reference grid;

(*b*) the size of the segment;

(*c*) the shape of the segment, here shown as the perimeter squared divided by the area;

(*d*) the smallest rectangle that can be placed around the entire segment (bounding rectangle).

In addition, the following are image spectral attributes calculated from the satellite image that are used in MICE processing:

(*a*) MEAN_CHANNEL_X, the mean grey-level value of channel X for the pixels in the segment;

(*b*) MAX_CHANNEL_X, the corresponding maximum pixel grey-level value of channel X in the segment;

(*c*) MIN_CHANNEL_X, the minimum pixel grey-level value of channel X in the segment.

These spectral values can be used to evaluate the spectral classification of the segment. These spectral attributes are by no means an exhaustive list for classification determination, but they do provide a basis upon which other attributes can be added. Thus, a simple classification rule could be as follows: if MEAN_CHANNEL_4 > MEAN_CHANNEL_2 and MEAN_CHANNEL_4 > MEAN_CHANNEL_1 then CLASS = LAND_COVER.

4. Instantiation and illustrative example

Knowledge for the MICE KBS is coded in the form of metarules, object rules and object values. The rules and values are stored in knowledge-base files that are input when an instantiated RESHELL expert is invoked. The MICE KBS uses metarules and object rules that are input from the knowledge base (long-term memory). The object values required for MICE processing are read in from the symbolic map and image files (short-term memory).

The role of the metarules is to establish the general procedure or strategy that is to be applied.

The object-level rules, on the other hand, contain the mundane details of the strategy that is being applied. As an example, there could be a metarule that states that the first action is to read in some file. At the object level, the corresponding rules could concern which portions of the file could be used.

RESHELL supports a hierarchy of intercommunicating expert systems. The advantage of this approach is that the problem can be decomposed into manageable portions, with limited interaction. For the map–image congruency evaluation expert the following decomposition, shown in scheme 1, was chosen. The role of each subexpert is as follows: (1) the map–image

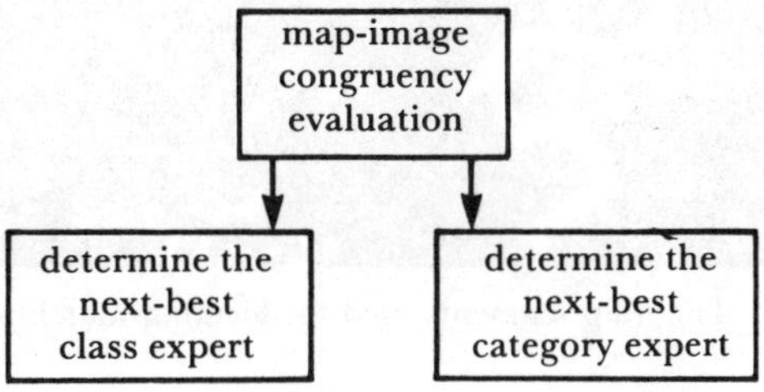

SCHEME 1. Hierarchical organization of the MICE KBS.

congruency evaluation expert is the highest-level expert controlling the input, processing and output of the KBS; (2) the next-best class expert returns the next-best class selected for congruency evaluation; (3) the next-best category expert returns the next-best category of the current class, selected for congruency evaluation.

The knowledge base required by the RESHELL architecture is such that one expert does not have access to the rules in another expert's knowledge base. In other words, the knowledge in the form of rules for each of the three MICE experts is separate and distinct.

Both the best class expert and the best category expert contain metarules and object rules that define the next best category, based on the current class and category. The class and categories were derived from the Canada Council on Surveys and Mapping list of categories of classes.

MICE has been tested by comparing the hydrography levels of a 1:20 000 scale BCMOFL forest cover map with the water features found in a *Landsat* MSS image (240 × 275 pixels). For this test, there were 25 metarules in the MICE expert, 20 object rules in the class expert, and 30 object rules in the category expert. The rules used are complex, with multiple conditions and actions. It takes approximately 3 h on a DEC AI VAX station to complete the knowledge-base phase of this comparison. The results were similar to those that would be obtained by a photograph-interpreter, in that MICE identified the same matches and mismatches between the image and the map. The human comparison of the map and the image was performed in less than an hour. The main rules coded into MICE were obtained through discussions with photograph-interpreters.

In the analysis of the spatial congruency, there were two image segments that matched a map segment for all four attributes; i.e. size, shape, class and overlap. These image segments matched map segments 3 and 9 of figure 4. There were an additional four image segments that matched 3 of the 4 possible attributes. These image segments matched map segments 1, 5, 6 and 11. Figure 4 shows the map segments and figure 5 shows the map segments that were selected by matching at least one attribute with its corresponding map segment.

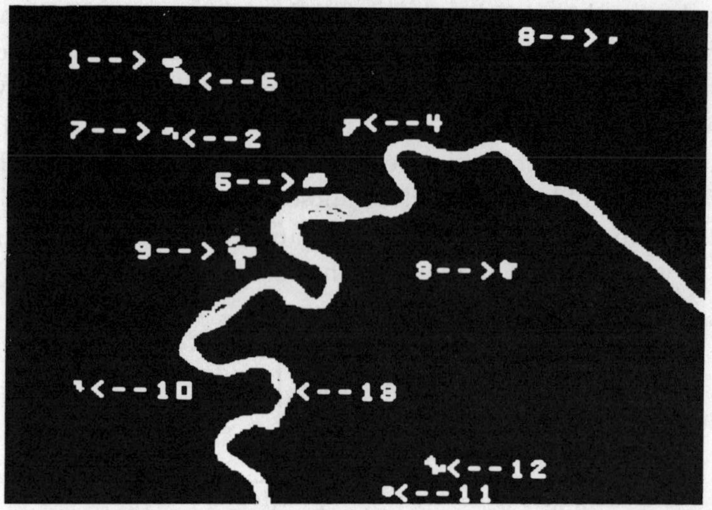

FIGURE 4. The map segments used for locating matching image segments.

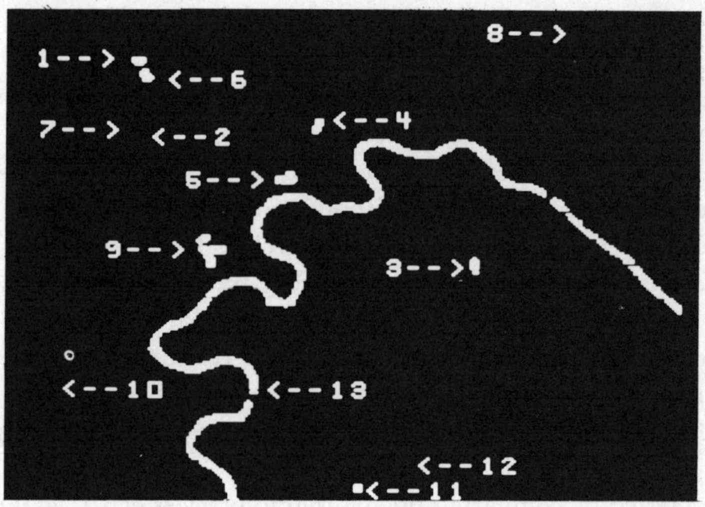

FIGURE 5. Image segments selected by matching a map segment by at least two attributes.

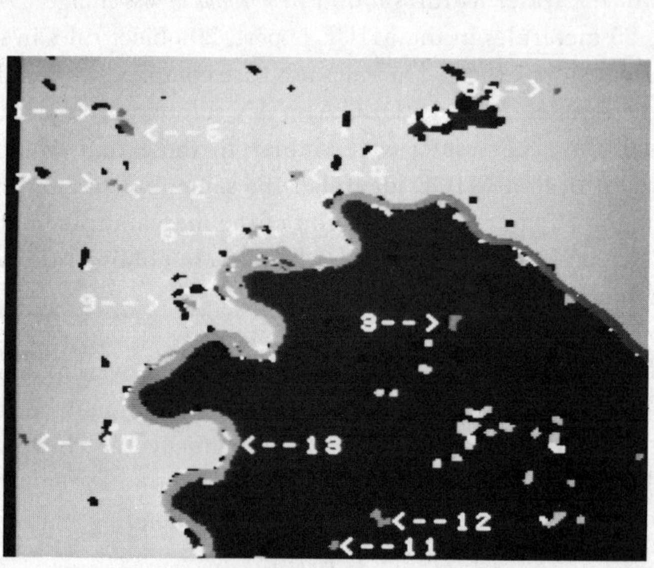

FIGURE 6. Congruency evaluation. The map segments are overlaid on the segmented image, showing areas of misregistration.

There remain 7 map segments that have no corresponding image segment. These segments are depicted in figure 4. Map segments 2, 7, 8, 10 and 12 were not matched to any image segment as the segmentation operator did not calculate any segments in those areas. Note that these are also quite small map segments. No match was found for map segments 4 and 13 because the segmentation operator generated multiple segments for the river and lake. These problems can be overcome by modifying the segmentation operator and by combining segments, but this is an area requiring further research.

Figure 6 shows the comparison of the congruency of the matching map and image segments. Note that there appear to be no systematic shifts in the map and image data; however, there is not perfect spatial congruency of the matching segments.

Conclusion

The results to date are quite encouraging for the use of a knowledge-based approach for the congruency evaluation of forest cover maps with remotely sensed imagery. In addition, the methods and techniques developed during the course of this research have applications in other processing of map and image data, such as map verification and image classification.

References

Billingsley, F. C. 1982 In *Proc. NASA Workshop on Registration and Rectification*, NASA-JPL Publication no. 82-23, pp. 13-20. California Institute of Technology.

Boyle, A. R. 1980 Scan digitization of cartographic data. In *Map data processing* (ed. H. Freeman & G. G. Pieroni); pp. 27-46.

Bryant, N. A. (ed.) 1982 In *Proc. NASA Workshop on Registration and Rectification*, NASA-JPL Publication no. 82-23, p. 517. California Institute of Technology.

Eshera, M. A. & Fu, K. S. 1986 An image understanding system using attributed symbolic representation and inexact graph-matching. In *IEEE Trans. Patt. Analys. Mach. Intell.* **PAMI-8** (5).

Glicksman, J. 1982 A cooperative scheme for image understanding using multiple sources of information. Ph.D. Thesis, University of British Columbia.

Goldberg, M., Goodenough, D. G., Alvo, M. & Karam, G. 1985 A hierarchical expert system for updating forestry maps with LANDSAT data. *Proc. IEEE* **73**, (6), 1054-1063.

Goodenough, D. G. 1988 The integration of remote sensing and geographic information systems. In *Remote sensing for resources development and environmental management* (ISPRS Commission 7) (ed. M. C. J. Darnen, G. Sicco Smit & H. Th. Verstappen), vol. 3. Rotterdam: A. A. Balkema. (In the press.)

Goodenough, D. G., Goldberg, M., Plunkett, G. W. & Zelek, J. 1987 An expert system for remote sensing. *IEEE Trans.* **GE-25**, 349-359.

Hayes-Roth, F., Waterman, D. A. & Lenat, D. B. 1983 *Building expert systems*. Reading, Massachusetts: Addison-Wesley.

Hegyi, F. & Sallaway, P. 1983 Integration of vector and grid data bases in B.C. forest inventory. In *Proc. 6th Int. Symp. on Automated Cartography* (ed. B. S. Wellar), vol. 1, pp. 215-221. The Steering Committee.

Marble, D. F. 1984 Geographical information systems: an overview. In *IEEE Pecora 9 Proceedings, Spatial Information Technologies for Remote Sensing Today and Tomorrow*, pp. 18-24.

McKeown, D. M. 1984 Knowledge-based aerial photo interpretation. *Photogrammetrica* **39**, 91-123.

McKeown, D. M., Harvey, W. A. & McDermott, J. 1985 Rule based interpretation of aerial imagery. *IEEE Trans. Patt. Analys. Mach. Intell.* **PAMI-7** (5), 570-585.

Parsons, T. J. 1984 Towards robust image matching algorithms, application of digital image processing VII. *SPIE* **504**, 436-444.

Plunkett, G. W. 1986 A map/image congruency evaluation knowledge based system. Master's Thesis, University of Ottawa.

Plunkett, G. W., Goodenough, D. G. & Goldberg, M. 1986 Map/image congruency evaluation knowledge based system. In *Proc. Graphics/Vision Interface, Vancouver, May 1986*, pp. 273-278

Winston, P. H. 1984 *Artificial intelligence*, 2nd edn. Reading, Massachusetts: Addison-Wesley.

Zarzychi, J. M. & Allam, M. M. 1982 Canadian Council on Surveys and Mapping – National Standards for the Exchange of Digital Topographic Data. Topographical Surveys Division, Surveys and Mapping Branch.

Discussion

D. LANE (*Intelligent Automation Laboratory, Department of Electrical and Electronic Engineering, Heriot-Watt University, Edinburgh, U.K.*). A well-known problem with blackboard-system architectures is the way in which knowledge sources (KS) may continually retrigger in a cyclic fashion. For instance, the contribution of KS A may trigger KS B, whose contribution triggers KS C. If the contribution of C retriggers A, then a cyclic condition exists. This effect is not always immediately obvious as it may involve large numbers of knowledge sources within one or a number of loops. How does your blackboard-system architecture deal with this?

M. GOLDBERG. In our blackboard system there is one master (expert) that is responsible for calling up (activating) the other knowledge sources (experts). It is up to this first expert to guarantee that no loops are formed.

Phil. Trans. R. Soc. Lond. A **324**, 457–465 (1988)
Printed in Great Britain

System architectures for interactive knowledge-based image-interpretation

By C. J. Taylor, J. Graham and D. Cooper

Department of Medical Biophysics, University of Manchester, Stopford Building, Oxford Road, Manchester M13 9PL, U.K.

[Plates 1 and 2]

We discuss hardware and software architecture for automated image-interpretation. The importance of considering the complete system is emphasized leading in particular to the conclusion that high-level and low-level processing are intimately linked. We present arguments to support the idea that automated image-interpretation systems should be knowledge-based and interactive. We attempt to identify the main architectural problems which such systems must address and outline a systematic strategy for acquiring, structuring and using knowledge.

Introduction

In this paper we consider the architectural issues raised by systems for automated image-interpretation. System architecture involves both hardware and software and we attempt to discuss their interrelation, arguing in the end that software problems are the more crucial. We begin by considering the broad characteristics of the image-interpretation task and identify high-level and low-level processing as qualitatively different activities. We review some of the specialized hardware configurations that have been developed to deal with the large quantities of data contained in images, highlighting the tendency for low-level processing to be considered in isolation. Against this background we discuss the desirability of a knowledge-based approach and suggest that interaction between the user and the system is also an important issue. We attempt to identify the central architectural problems and argue that they are not amenable to solution simply by the application of massive computing power but rather are truly architectural in nature and require a strategy for selecting, structuring and using information. Finally we propose a systematic approach to some of the fundamental problems that have been identified. We describe this as a software architecture to emphasize that architecture is not concerned solely with hardware. The architecture makes use of explicit image models and is the subject of current research.

Where possible we have used, as illustrations, practical examples from our own work in medical and industrial image-interpretation. We believe, however, that the ideas are perfectly general and have attempted to relate them to remotely sensed imagery where possible.

The first illustrative problem is chromosome analysis. Figure 1*a* shows the genetic material from a single human cell arranged, as it is during cell division, into bodies of characteristic size and shape called chromosomes. When such images are used clinically to detect genetic abnormalities, a technician uniquely identifies each of the individual chromosomes by its size, shape and pattern of stain uptake (banding pattern). They are normally arranged in a regular display called a karyogram (figure 1*d*). A machine that could take such cells and automatically generate a karyogram would be ideal. In practice, a semi-automated, interactive system is clinically useful (Graham 1988).

The second illustrative problem is automated industrial inspection, specifically the inspection of motor-car drum-brake assemblies (figure 2). In this case the task is to recognize and locate each of the component parts (which are moveable) and check that they are present, correctly fitted and undamaged (Woods *et al.* 1987).

Levels of processing

It is convenient to identify two qualitatively different types of processing that are required for automated image-interpretation. Low-level processing is primarily numerical and acts directly on image data. Examples are geometric correction, stereo ranging, homogeneous region extraction and edge-detection. In general, low-level processing takes one or more images as input and produces a different image as output. Typically, the result at a point in the output image depends on the intensity values in a small neighbourhood surrounding the corresponding point in the input image. An example is edge-detection which is shown in figure 2*b*, *c*.

High-level processing is largely symbolic and involves recognizing and describing image structures. Typically this involves matching the observed data to some model of the expected appearance of known structures. For instance, for the brake assembly we might store an idealized edge map for each subcomponent and attempt to match these to the edges detected in an observed image. In remote sensing a comparable example might involve the use of map data as a model to which observed data must be related.

High-level and low-level processing present very different characteristics. Low-level processing typically involves very simple repetitive operations performed on large data sets, whereas high-level processing is essentially symbolic and involves complex processing on relatively small data sets. It is important to recognize that both types of processing are involved in automated image-interpretation and to reflect this in system architecture.

Specialized hardware

In this section we consider some of the hardware arrangements that have been developed to address specific problems posed by image-processing. Much of the effort has concentrated on low-level processing where significant computing power must be brought to bear, though we argue later that these problems should not be considered in isolation.

Coprocessor systems

Figure 3*a* shows the arrangement of a coprocessor system (see, for example, Taylor *et al.* 1986). In such a system a relatively small number of specialized processors co-operate to act upon image data and other types of data held in memory. Operations such as data-processing, memory-address calculation and program-flow management can be performed simultaneously by processors specializing in these activities. The coprocessors can be designed and programmed to make selected primitive operations very efficient and can thus achieve realistically high performance for low-level processing. Such an arrangement is, however, extremely flexible and can also be designed to perform high-level processing efficiently within the same structure.

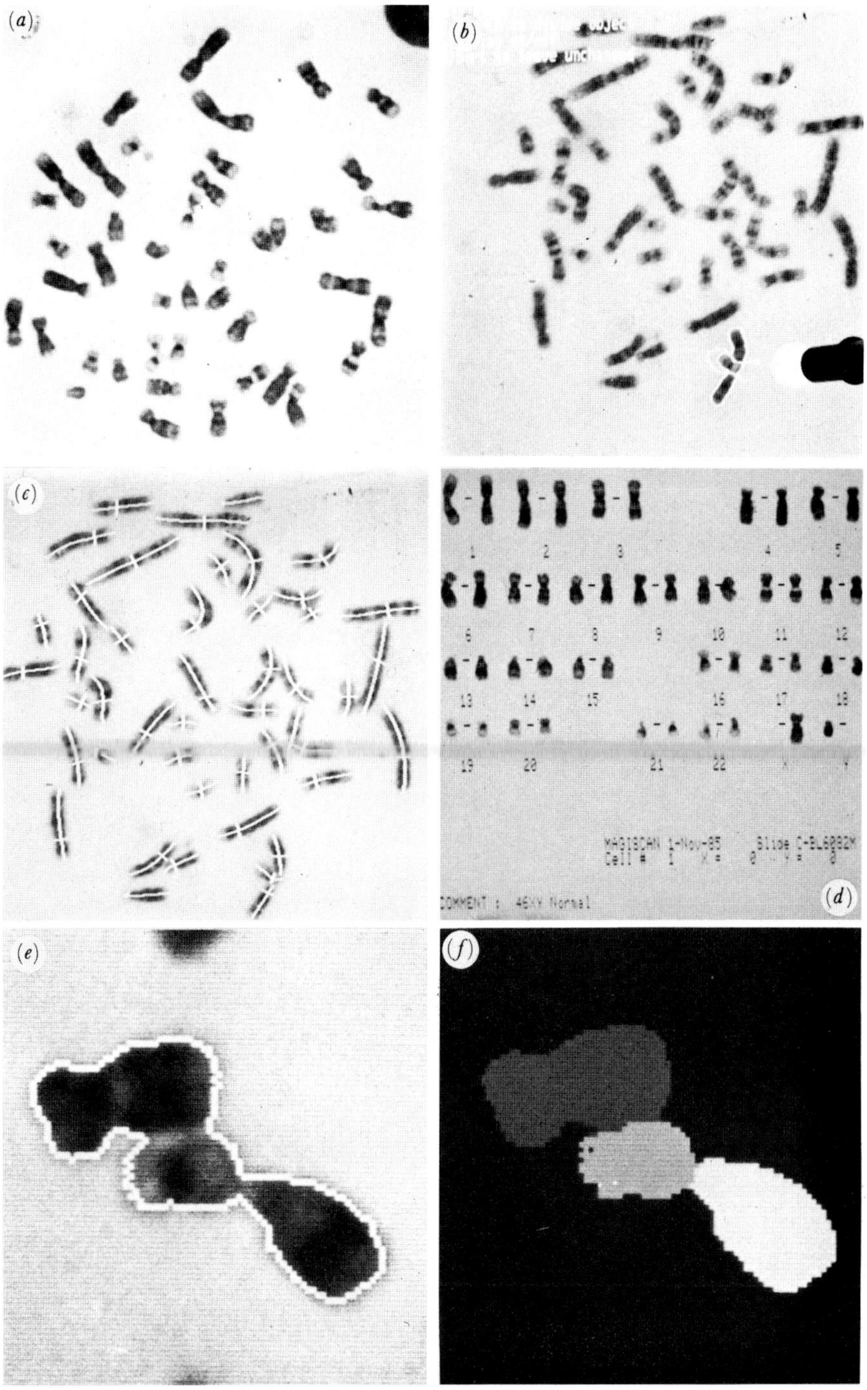

FIGURE 1. Chromosome analysis. (*a*) Microscope image of a dividing human cell; (*b*) user interacting with an automated analysis system to separate overlapping chromosomes; (*c*) axis and centromere automatically located for each chromosome; (*d*) automatically generated karyogram; (*e*) erroneous initial hypothesis for a chromosome boundary; (*f*) modified hypothesis generated as a result of obtaining further low-level evidence.

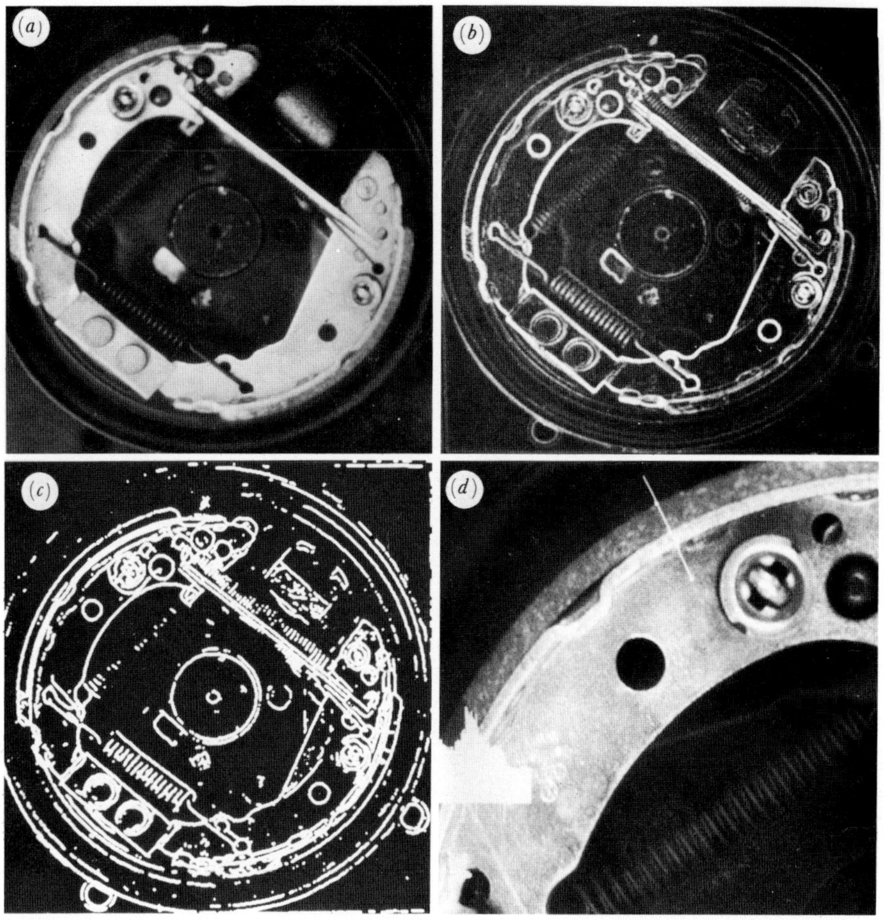

FIGURE 2. Brake inspection. (*a*) Plan view of a motor-car drum brake assembly; (*b*) edge strength image of (*a*); (*c*) detected edges from (*b*); (*d*) line which defines the position at which brake lining thickness should be measured. The intensity profile along this line is displayed and markers on the line indicate the points between which the system intends to make the measurement.

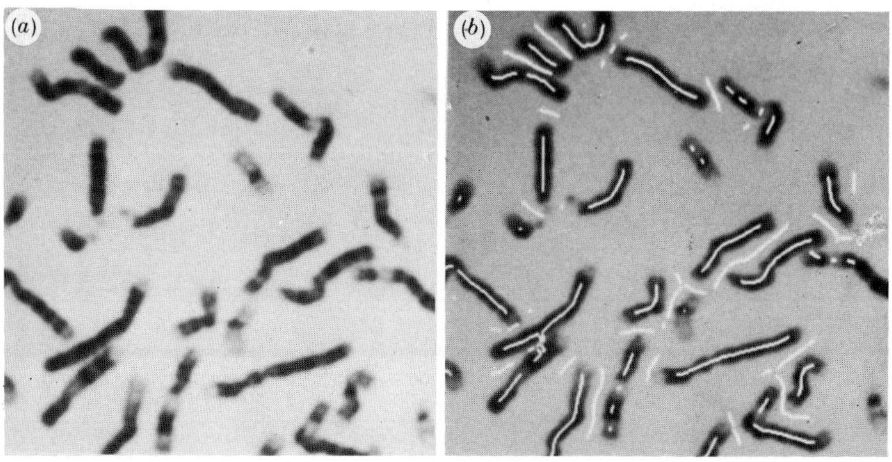

FIGURE 6. Centre of symmetry cues. (*a*) Original image-containing chromosomes; (*b*) loci of intensity symmetry detected directly from the intensity image.

Pipeline processors

A simple arrangement, which achieves high performance for low-level processing tasks, is shown in figure 3*b* (see, for example, Gerritsen & Monhemius 1981). The pipeline processor makes use of the fact that in low-level processing an identical sequence of operations must often be performed on a large number of pixel values. Each processor in the pipeline undertakes one step in the sequence of operations. Each pixel value is sent down the line in turn so that if there are n stages in the pipeline n pixels are operated on at once. If all the processors have the same speed then the system is n times as fast as an individual processor. Once high-level processing is involved the arrangement offers virtually no assistance.

Array processors

The arrangement of an array processor is shown in figure 3*c* (see, for example, Duff, 1979). These systems make use of the characteristic of many low-level algorithms that they involve operations on small neighbourhoods of pixels. In this radically different arrangement there is a processor and memory element for each pixel. Each processor–memory element is connected to each of its neighbours so that it can act upon its own and neighbouring pixel values. Thus by sending the same instructions to each processor an operation such as edge-detection can be performed for each pixel in the image simultaneously. Again, although this configuration is well suited to low-level processing it is inappropriate for high-level processing because the connections between processors are insufficiently general.

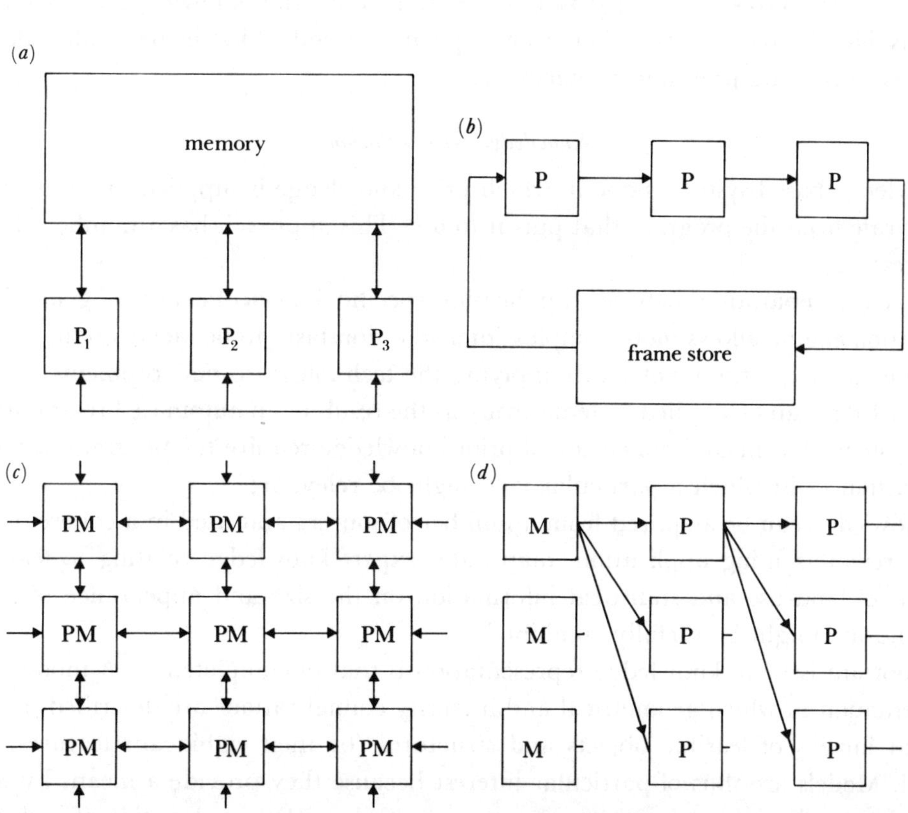

FIGURE 3. Specialized hardware for image processing. (*a*) Coprocessor system; (*b*) pipeline processor; (*c*) array processor; (*d*) adaptive network. P, processor; P_i, specialized processor; M, memory.

Adaptive networks

Adaptive networks (figure 3*d*) bear a superficial resemblance to the organization of the brain and represent a more radical architectural approach (see, for example, Hopfield 1982). Here the processors only ever execute one type of operation, combining the set of inputs according to a predetermined rule to produce an output. The nature of the processing that takes place is thus determined not by the processors but by the pattern and strength of the connections between processors. In principle, the final output of such a system can be either another image or a symbolic interpretation. Adaptive networks are very interesting but so far no-one knows much about how to set up the connections to achieve the desired result or indeed about their generality. A particular concern is that knowledge of the problem domain must necessarily be incorporated implicitly into such systems, a characteristic which, as we argue in the next section, is undesirable.

Interactive knowledge-based interpretation

Images from a particular domain, such as remote sensing, are normally interpreted by making use of specialized knowledge. This knowledge includes the nature of the interpretation task together with the identity and expected appearance of key objects and structures. Automated systems have been developed to interpret complex images (Aleksander 1983; Brunt *et al.* 1983; Dixon & Taylor 1979; Graham 1987; Pycock & Taylor 1980; Thomason 1986; Tucker & Shippey 1983; Woods *et al.* 1987) but generally the prior knowledge involved has not been easily identifiable; it has rather been implicit and embedded in particular algorithmic approaches to the interpretation problem.

Knowledge representation

A knowledge-based system is one in which prior knowledge is supplied in an explicit form, quite separate from the program that puts it to use. This approach has a number of potential advantages:

(i) the clear separation between application-specific knowledge and a general-purpose interpretation engine allows more complex interpretation tasks to be tackled reducing the cost and engineering expertise involved in applying the technology to new problems;

(ii) knowledge can be applied systematically as the result of an automated reasoning process whereas systems that make implicit use of prior knowledge require the programmer to foresee the circumstances in which a particular fact might be relevant;

(iii) knowledge can be acquired from a number of sources and used in a coherent manner; thus in a remote-sensing application, map data, expert knowledge relating to feasible configurations of land-use and statistical information on the size and appearance of regions of known land-use might be usefully combined.

An important issue in knowledge representation is that of completeness. A model is a form of representation in which geometrical and intensity configurations are described in sufficient detail that images of feasible objects and structures (or their significant features) may be generated. Models are thus of particular interest because they provide a means by which an automated image interpretation system may arrive at a complete explanation of each observed image (Ayache & Faugeras 1986; Brooks 1981; Hanson & Riseman 1978; Pollard *et al.* 1987).

In addition to a method of knowledge representation a practical system requires a means by which the user can supply relevant knowledge. Although some knowledge can be presented to the system formally (in a manner analogous to conventional programming) there is also a need for interactive dialogue between the user and the machine. Two types of interaction may be considered. The fact that both involve the user gaining rather direct access to the geometrical and intensity structure of an image provides additional support for a model-based knowledge representation that can form a natural link between the user interface and the internal-processing régime.

Offline interaction

Although some knowledge can be stated simply, a great deal of what is used to interpret images can only be conveyed conveniently by showing examples. This is indeed the method we often use when explaining a complex visual task to another person. The requirement is that the user can present example images and interact with the system in such a way that it extracts salient features. We show this with an example from the interpretation of brake assembly images.

In figure 2*d* the system has identified the position at which the thickness of the brake lining is to be measured. The intensity profile observed along the measuring line does not, however, allow the inner and outer edges of the lining to be easily identified. The figure shows the system making an educated guess at the location of the inner and outer boundaries. The user accepts the guess if it is accurate or corrects it with a lightpen if it is erroneous. This is repeated for a number of examples and because, in each case, the system has both the observed intensity profile and the correct interpretation it can, by building an internal model, learn how to guess correctly so that no interaction is required at run-time.

Online interaction

In many circumstances it may be desirable for the user to provide additional knowledge at run-time. This may be appropriate because complete automation is too difficult or because a user-guided system is more appropriate for the task in hand. We show this with an example from chromosome analysis (figure 1). Here the generation of a karyogram can be significantly automated but occasionally configurations arise that are difficult for the system to resolve. In any case, because the system is used diagnostically, skilled supervision is desirable. The system starts by determining a boundary for each chromosome. Figure 1*b* shows how the user can separate overlapping chromosomes with the lightpen. To decide which chromosome is which, an axis of symmetry must be obtained for each and used to define a path along which the banding pattern should be measured and the position of the centromere located. Figure 1*c* shows a display of the results, which can again be corrected with the lightpen. Finally the chromosomes are classified and displayed in a karyogram, which can be modified if necessary by moving chromosomes with the lightpen (figure 1*d*).

The architectural problem

Having identified some of the important characteristics of an ideal image-interpretation system we can consider some of the architectural issues which are raised. First let us look at the question of computational complexity. It is often argued (Sternberg 1980) that this is a particularly important issue in low-level processing because of the large quantities of data involved. Figure 4 shows with a simple example that there is also a problem with high-level processing. The figure shows a line-drawing model of a house, composed of straight-line segments and an example image from which we may assume all the line segments can be perfectly recovered. To recognize the house we must match the model and the example line by line. The number of ways to do this goes as the factorial of the number of line segments and in this example there are approximately 10^{20} ways of matching. To recognize the house in one second by using brute force would require computing power of at least 10^{14} Mips (million instructions per second). To put this in perspective NASA's massively parallel processor provides approximately 10^4 Mips (Batcher 1980). Thus it is unrealistic to assume that the application of parallelism will solve the problem, neither can it be avoided because any of the matches might be the best one. What is required is a framework within which the system can focus its attention on likely interpretations avoiding computational effort being wasted on those which are unlikely.

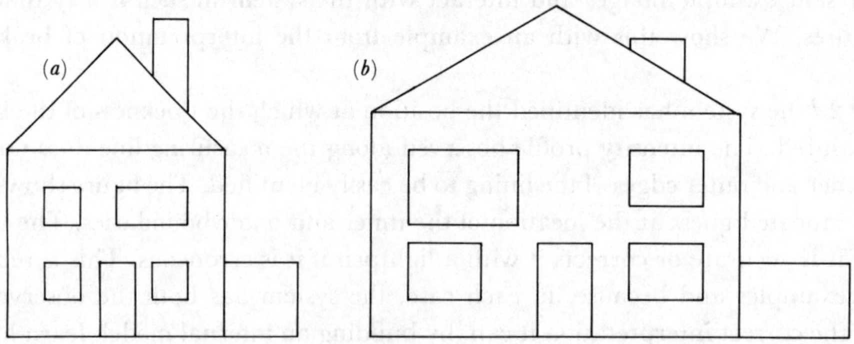

FIGURE 4. (*a*) Idealized house model composed of straight line segments; (*b*) the line segments which might be obtained for an example of a house image.

To appreciate fully the nature of the architectural problem with which we are faced we must address a further complication. In our discussion of model matching we have assumed that low-level processing may be used to extract evidence of structure (e.g. line segments) which may subsequently be interpreted by matching to a model. In practice this is unrealistic because the evidence of structure obtained by unguided low-level processing is subject to errors and will not in general represent the best evidence that could be sought in support of an emerging interpretation hypothesis. A more realistic approach is to undertake some low-level processing to provide sufficient evidence to make an initial interpretation which can then be used as an organizing hypothesis to guide the collection and interpretation of further low-level evidence. This further evidence may lead to the rejection, modification or refinement of the original hypothesis. Figure 1 *e*, *f* show an example in chromosome analysis where initial low-level processing identifies a composite object. From its shape this is recognized as unlikely to be a single chromosome and further low-level evidence is sought which generates a modified hypothesis of three objects (Graham *et al.* 1986).

In summary there are two main conclusions which we draw from this discussion:

(i) high-level and low-level processing are intimately linked and the interface between them is crucial;

(ii) real visual tasks involve massive computational complexity and the central issue is that of deploying the available computing resources to best effect rather than increasing them.

We (and others) have designed coprocessor hardware which optimizes the interface between different levels of processing (Taylor *et al.* 1986; Graham *et al.* 1986) and have investigated strategies for focusing computing resources in a number of problem domains (Dixon & Taylor 1979; Brunt *et al.* 1983; Pycock & Taylor 1980; Woods *et al.* 1987; Graham 1988).

A SOFTWARE ARCHITECTURE

In previous sections we have argued that the problem of image-interpretation should be posed as that of explaining an observed image in terms of an explicit image model. The model will describe the objects and structures that may appear in the image and the relations between them. It will be parametrized so that a particular set of parameter values uniquely defines the geometric and (ideally) intensity configuration of a particular feasible image. Typical parameter values and measures of variability will have been obtained by offline interaction with example images. Image-interpretation involves a search for that set of model parameters that are consistent with those observed in example images and that define an image most similar to the observed image.

Hierarchical models

We showed in the previous section that the cost of establishing a correspondence between model elements and image structures rises exponentially with the number of model elements. To model visual worlds of realistic complexity requires that the matching problem be divided into a number of subproblems of manageable cost. This can be achieved in a natural way by organizing the model as a hierarchy. Figure 5 shows the manner in which the house model of figure 4 might be broken down into a number of submodels. In this example a partition of the model such as BODY will contain information describing the expected geometrical relations between submodels such as WINDOW and WALLS. Each submodel at the lowest level describes the relations between the line segments of which it is composed. If we can match each submodel independently the number of segment to segment matches which must be considered is reduced from approximately 10^{20} to approximately 10^3.

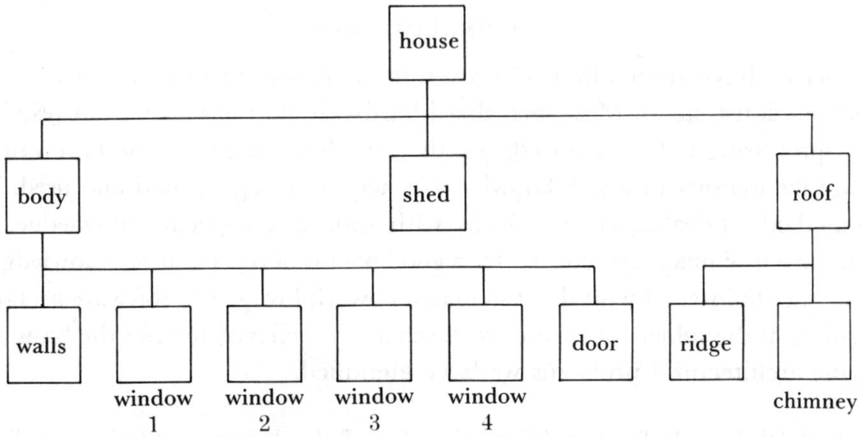

FIGURE 5. The house model of figure 4 partitioned into a hierarchy of submodels.

Cues and control

A cue is a feature that suggests that a particular model element should be instantiated. Given a hierarchially organized model and an observed image it is generally necessary to apply low-level processing to generate cues that provide evidence with which to initiate the matching process. In the house example we might detect intensity discontinuities (edges) in the observed image and starting from the bottom work upwards. Once a submodel has been matched its relation to other submodels can be used to limit the search space. For example, once WALLS has been recognized the approximate position of each WINDOW is defined and may be used as a cue which propagates the matching process.

It is important to recognize that it is not necessary to start matching at the lowest level of the model. In the house example we might, for instance, use a cue-generator that found dark blobs of about the size of a DOOR. On finding such a cue the system would try to recognize and locate the rectangle of the DOOR perhaps seeking edge information in the vicinity of the cue. If the DOOR were located the system would move up a level in the hierarchy and use the known relation between DOOR and WALLS to generate a WALLS cue. If the WALLS were successfully located then the WINDOWS could be cued. This process of cue-driven matching can continue until a complete match has been established. An important feature of this organization is that it is possible for the system both to infer higher-level models from details and details from higher-level models. Ultimately, however, the interpretation must be supported by direct evidence. It is also important to note that the pattern of control is determined, as seems appropriate, by the nature of the images to be interpreted.

Figure 6, plate 2 shows a practical example of a cue generator which operates robustly at a higher level than edges. The figure shows a chromosome image from which loci of intensity symmetry have been extracted directly without first detecting edges or separating objects from background. These lines represent candidate chromosome axes which can be refined and tested by appealing to other levels of a model to recognize chromosome boundaries, banding pattern and so on. Cues such as this which detect reasonably high levels of organization in the image are important because they will, as a result, tend to be less ambiguous.

CONCLUSIONS

In the paper we have argued that software is more important than hardware in automated image-interpretation systems. Most specialized hardware that has been built provides support for low-level processing but does not address the complete image-interpretation problem. The real issues are the manner in which knowledge is acquired, represented and used. The need is to develop methods of dealing systematically with application-specific knowledge. We suggest that explicit models of image structure offer a good means of representing knowledge internally and allow the user to interact with the system in a powerful way. The software architecture that we have outlined is the subject of current research and is believed to offer the basis of a solution to some of the architectural problems we have identified.

We thank P. W. Woods for providing pictures of the brake inspection application, and S. A. Thornham and P. J. Azzopardi for their assistance in preparing other figures.

References

Aleksander, I. 1983 *Patt. Recog. Lett.* **1**, 375–384.

Ayache, N. & Faugeras, O. D. 1986 *IEEE Trans.* **PAMI 8**, 44–54.

Batcher, K. 1980 *IEEE Trans Comput.* **29**, 836–840.

Brooks, R. A. 1981 *Artif. Intell.* **17**, 285–348.

Brunt, J. N. H., Taylor, C. J. & Dixon, R. N. 1983 In *Physical techniques in cardiological imaging* (ed. M. D. Short), pp. 153–162. Bristol: Hilger.

Dixon, R. N. & Taylor, C. J. 1979 In *Machine aided image analysis, 1978* (*IOP Conference Series no.* 44) (ed. W. E. Gardner), pp. 178–185.

Duff, M. J. B. 1979 In *Advances in digital image processing* (ed. P. Stucki), pp. 265–276. New York: Plenum Press.

Gerritsen, F. A. & Monhemius, R. D. 1981 In *Languages and architectures for image processing* (ed. M. J. B. Duff & S. Levialdi), pp. 189–203. London: Academic Press.

Graham, J. 1988 *Analyt. Quant. Cytol. Histol.* (In the press.)

Graham, J., Taylor, C. J., Cooper, D. H. & Dixon, R. N. 1986 *Patt. Recog. Lett.* **4**, 325–333.

Hanson, A. R. & Riseman, E. M. 1978 In *Computer vision systems* (ed. A. R. Hanson & E. M. Riseman), pp. 303–333. Orlando, Florida: Academic Press.

Hopfield, J. J. 1982 *Proc. Natn. Acad. Sci. U.S.A.* **79**, 2554–2558.

Pollard, S. B., Porrill, J., Mayhew, J. E. W. & Frisby, J. P. 1987 *Image Vision Comput.* **5**, 73–78.

Pycock, D. & Taylor, C. J. 1980 *Analyt. Quant. Cytol.* **2**, 195–202.

Sternberg, S. R. 1980 In *Real-time medical image processing* (ed. M. Onoe, K. Preston & A. Rosenfeld), pp. 11–22. New York: Plenum.

Taylor, C. J., Dixon, R. N., Gregory, P. J. & Graham, J. 1986 In *Intermediate-level image processing* (ed. M. J. B. Duff), pp. 19–34. London: Academic Press.

Thomason, R. L. 1986 In *Vision Conference Proceedings, Detroit, 1986*, pp. 5:51–5:61.

Tucker, J. H. & Shippey, G. 1983 *Analyt. Quant. Cytol.* **5**, 129–137.

Woods, P. W., Taylor, C. J., Cooper, D. H. & Dixon, R. N. 1987 *Patt. Recog. Lett.* **5**, 11–17.

Discussion

D. Lane (*Intelligent Automation Laboratory, Department of Electrical and Electronic Engineering, Heriot-Watt University, Edinburgh, U.K.*). In Dr Taylor's presentation he mentioned the subject of feedback, and described the way that high- and low-level processes may interact. Feedback and the ensuing issue of system stability have been much studied by mathematicians and control engineers for a number of years. Mathematical tools (pole-zero diagrams, nyquist plots, bode diagrams) have been developed to enable a designer to predict the stability margins of a system. Is the stability issue relevant in the context of knowledge-based image-interpretation, and if so, how may it be approached?

C. J. Taylor. The issue of stability is relevant, but the system with which we are dealing is much more complicated than those for which the mathematical tools mentioned were developed. Feedback is involved in knowledge-based image interpretation in the sense that high-level hypotheses may be used to guide the search for low-level supporting evidence which in turn may be used to modify the high-level hypotheses; a high-level interpretation is stable but it is not necessarily so that such a system will converge to that interpretation. The interaction between high-level and low-level processing is, however, sufficiently complex that it is difficult to see how the methods of control engineering can easily be applied.

Phil. Trans. R. Soc. Lond. A **324**, 467–469 (1988)
Printed in Great Britain

General discussion

SIR HERMANN BONDI, K.C.B., F.R.S. (*Churchill College, Cambridge, U.K.*).

(i) I see that our main concern is to stimulate the potential civil users to come forward and let us know their requirements with some precision. Because it has become clear that there are no general answers to all questions, our work needs to be targetted. The military user can be more definite and say that he wants to see where specific targets are. Can one be equally specific about particular civil needs?

(ii) Nor do I believe that money need be an obstacle. In the commodity markets an interpretation of remotely sensed data ahead of other knowledge should give a tremendous advantage to somebody buying or selling futures in, say, coffee. For, with sufficient knowledge and expertise, the data should reveal the state of the crop in all parts of the world.

How can these sources of funds be tapped?

(iii) What other civil uses might involve very large sums of money?

J.-P. A. L. MULLER.

(i) Yes. Examples include the state of the atmosphere and oceans; where changes in land use or other anthropomorphic activities have occurred (in, for example, the revision of civilian maps or the monitoring of the Chernobyl nuclear disaster) to the location of mineral resources.

(ii) The problem is that remotely sensed data cannot yet be used to provide the information with a sufficient confidence and to a given level of accuracy required by the commodity markets. The principal difficulties are in the sensor designs and the reflectance extraction algorithms which cannot be used to identify a unique spectral signature. Hope remains with steerable imaging spectrometers which will enable more quantitative information to be extracted on the surface cover by both looking for key factors (e.g. at a particular bi-directional view for certain wavelength combinations) and removing atmospheric effects.

(iii) Oil and gas exploration, and national mapping, might involve very large sums of money.

C. J. OLIVER. Research into SAR image-interpretation is unfortunately expensive to embark upon in terms of both image-production and image-interpretation. It is therefore one of those chicken and egg situations in that you are most likely to get funds for research you have already been successful in! However, the military user may have tasks of such high military priority that he is prepared to fund high-risk research. In the civil field, though the same principle should apply, the financial balance between cost and possible effectiveness is probable more difficult to quantify. Various lessons from MoD research may, however, be relevent to the civil field. For example, it is not productive to embark initially on a global image-understanding approach because the problems are so wide ranging. Instead one should endeavour to factorize out from the total problem various key elements, each with important consequences, which may be tractable within a reasonable time frame. Such clearly defined tasks can be convincingly demonstrated to succeed (or fail) and will tend to establish a climate in which more speculative ventures, with possibly greater potential benefits, may be supported. Expensive long-term research should be associated with shorter-term real applications arising from the total research programme.

Sir Hermann Bondi, K.C.B., F.R.S. Many years ago already, remotely sensed pictures were splendid wall decorations for the managing director's office. Today we learned of the long and laborious path from the cosmetically pleasant to the useful. It is good to hear that there are people putting in this hard and prolonged effort. However, to focus it well, we need the users. When do contributors believe the work will be sufficiently far advanced to stimulate our prospective customers?

J.-P. A. L. Muller. I believe that this will happen when both unique questions can be formulated and the problems of multilevel knowledge invocation can be solved, particularly low-level feature extraction.

C. J. Oliver. My presentation on aspects of clutter modelling and superresolution concentrated on some of the more speculative aspects of our research programme, aimed about 10 years ahead. Within the same reasearch section we have produced important patented applications on the automatic production of focused, distortion-free, SAR images enabling detailed comparison of different images to be made. Research on these capabilities is essentially complete and we are in the development stage in which the methods are being installed on the RSRE SAR processor. Thus, research into a few important problems is already sufficiently far advanced for exploitation to be seriously considered. One would expect that continuing long-term research should progressively yield further capabilities.

S. Quegan. The path from the pleasant to the useful is longer and more laborious than it should be, largely because the funding and effort put into interpretation of data is in many cases grossly out of proportion to the large amounts expended in acquiring that data. The results presented have been the fruit of a small group of people who are taking the interpretation problem seriously.

As far as SAR is concerned, users need to have data to be interested, and until the launch of *ERS*-1 such data will not be readily available. It is essential, however, that along with the data we supply the interpretation tools. At the present level of effort, I do not believe that these tools will be ready in time for *ERS*-1 (at least as regards land use), and unless the effort is expanded it would be optimistic to expect them before the end of the century.

J. R. Milford (*Department of Meteorology, University of Reading, U.K.*). I would like to draw attention to an example from our work, funded by the U.K. Overseas Development Administration, which illustrates the Chairman's opening remarks particularly well.

We are developing methods for monitoring tropical rainfall by using *Meteosat* data: even if this can only be done crudely it will be useful in data-sparse regions, where there is no viable alternative for estimating the average rainfall. The intermediate product is the total duration of cold cloud (temperature below −60 °C) over a ten-day period. Our research aims to find the optimum relation between the duration and the rainfall.

It is clear that any regression will leave a sizeable residual variance, but comparison of cross sections through the cloud duration map and the isohyetal map drawn from surface gauges shows close qualitative correspondence, particularly in the positions of the minima. It may be that this alone makes the procedure worthwhile because rainfall minima over two or more succesive ten-day periods during the rainy season would immediately indicate areas where most forms of agriculture are at risk.

To return to the Chairman's remarks, our experience shows the considerable effort which is needed to develop data-processing methods to the stage at which they match the sophistication of the data-gathering hardware. Development of the rain estimating technique has required about 20 man years, and will need as much again before we have operational methodology with known limits to the accuracy of the outputs in different regions of Africa. Fortunately specialist hardware and software is under development within the UN FAO.

The second point which we would support is that we need better communications with the users, such as FAO, UNDP, regional organizations and governments to match their specific needs with the type and reliability of outputs which we can deliver.

A third point is that the outputs are not of a kind which can bring quantifiable commercial rewards, and therefore have any identifiable market price. An international long-term programme is the only feasible kind of support for the satellite and the data processing system, and we are fortunate to be working with *Meteosat*, to which Europe already has a long-term commitment.

Phil. Trans. R. Soc. Lond. A **324**, 471–472 (1988)
Printed in Great Britain

Concluding comments

By D. Stanley

Organisation and System Innovations Limited, Tectonic Place, Holyport Road, Maidenhead, Berkshire SL6 2ET, U.K.

This has been a most stimulating meeting, not least of all because of the healthy mix of people attending. We have heard lively discussions between users and researchers, both in the formal sessions and during the breaks. Representation from widely diverse application areas and examples taken from a variety of images demonstrate that we represent a unified community of interest.

Many of us sense that this meeting has occurred at a watershed in the development of image-analysis as applied to applications as demanding as remote sensing. With an acute awareness of the limitations of pixel-based image-processing, it is easy to see the appeal of knowledge-based techniques that promise to build in some of the richness of human capability. If this can be achieved without including human limitations, such as the tendency to subjectivity and increasing error rates during repetitive operational tasks, then so much the better.

However, despite the emphasis of this meeting on operational applications, we have seen during the second day that the breakthrough may still be some way off in terms of 'intelligent' applications. This makes it difficult to predict future trends accurately, but there are a number of reasons why the subject has been slow to take off in terms of large-scale operational exploitation, not all of them technical.

I suspect that one of these barriers is that R & D is often not linked closely enough to the requirements of operational users. This creates a credibility gap, arising from doubts about cost effectiveness, reliability, performance, support and user friendliness. Another key organizational problem that must be overcome is associated with the management of knowledge itself. Knowledge is still power in many hierarchies and it may prove difficult to find experts willing to share their hard-won expertise. This may explain the preponderance of user-assistance rather than automatic applications in the presentations.

Given the rapidly increasing quantity of data available, however, we must expect to see a continuing trend towards more autonomous, real-time interpretation systems. Fortunately, the trends in computer power, with an order of magnitude improvement in cost–performance every five years, is in our favour. Developments in parallel processing will make knowledge-based image-understanding systems a reality within a few years. Already we are in an era where hardware is no longer the problem. The limiting factor is now our ability to organize information and handle software complexity. This, of course, is actually a management problem. Useful applications of remotely sensed data will require integration with other strategic management information.

Perhaps one of the most promising benefits of knowledge-based techniques will be the increase in flexibility. Modification of algorithm-based analysis systems is notoriously cumbersome and expensive, whereas expert systems can grow with the increasing experience and sophistication of the user. They also will correct the tendency of more rigid classification systems to encourage the interpreter to close off options prematurely. Nevertheless, they suffer acutely from the problem of all computer systems that if you put garbage in, you will only get garbage out.

We should be grateful to all those who have worked to make this meeting such a success, but especially to Dr B. J. Conway of the Meteorological Office and Christine Johnson of the Royal Society.